Edward S. Sears

Faxon's Illusrated Hand-Book of Summer Travel

Edward S. Sears

Faxon's Illusrated Hand-Book of Summer Travel

ISBN/EAN: 9783337211608

Printed in Europe, USA, Canada, Australia, Japan

Cover: Foto ©Andreas Hilbeck / pixelio.de

More available books at **www.hansebooks.com**

FAXON'S
Illustrated Hand-Book

OF

SUMMER TRAVEL,

TO THE

LAKES, SPRINGS AND MOUNTAINS

OF

NEW ENGLAND AND NEW YORK.

BY EDWARD S. SEARS.

ROUTES TO NEW YORK, LONG ISLAND SOUND, THE HUDSON RIVER THE HOOSAC TUNNEL AND SARATOGA. LAKES GEORGE, CHAMPLAIN, WINNEPISEOGEE AND MEMPHREMAGOG. THE WHITE MOUNTAINS, AND THE ADIRONDACKS. NIAGARA, THE ST. LAWRENCE AND THE SAGUENAY, ETC., ETC.

NEW AND REVISED EDITION

BOSTON:
PUBLISHED BY CHARLES A. FAXON,
NO. 228 WASHINGTON STREET.
(FORMERLY 82.)
1875.

Contents.

CHAPTER I.

STEAMBOAT ROUTES TO NEW YORK. — Stonington Line; Providence; Steamer Rhode Island; Approach to New York; The Fall River Route; Newport; Norwich Line; Norwich; Shore Line Route to New York; New Haven; Springfield Route to New York; Hartford and the Connecticut River.

CHAPTER II.

UP THE HUDSON RIVER TO ALBANY. — Saratoga; The Palisades; Highlands; West Point; Newburgh; Fishkill; Poughkeepsie; Hudson City; Albany; other Routes to Saratoga; Fitchburg. Rutland and Saratoga Line; Fitchburg; Mount Monadnock; Bellows Falls; Ludlow; Rutland; Rutland to Saratoga; Boston and Albany Route; Worcester; Springfield; Pittsfield to Albany.

CHAPTER III.

THE HOOSAC TUNNEL ROUTE TO SARATOGA. — Fitchburg; Gardiner; Athol; Orange; Brattleboro'; Greenfield; Hoosac Tunnel and its History; North Adams to Troy.

CHAPTER IV.

SARATOGA AND ITS ATTRACTIONS. — Congress Park; Congress, Columbian, Empire, Star, Pavilion, Hathorn and Geyser Springs; Analysis of the Waters; Bottling the Waters; United States, Grand Union, Congress and

CONTENTS.

Columbian Hotels; The Holden and Waverly Houses; Strong's Institute; Social Life in Saratoga; Schroon Lake; The Route to Lake George.

CHAPTER V.

LAKE GEORGE AND ITS BEAUTIES. — The Fort William Henry Hotel; Trip down the Lake; History of Lake George; Ticonderoga.

CHAPTER VI.

DOWN LAKE CHAMPLAIN. — Crown Point; Burlington; Lake View; Across the Lake to Plattsburg; Scene of the Battle, etc.

CHAPTER VII.

PLATTSBURG AND THE AUSABLE CHASM. — Fouquet's Hotel; Plattsburg; The Drive to the Chasm; Map of the Chasm; The Journey through the Glen, etc.

CHAPTER VIII.

THE ADIRONDACK REGION. — Paul Smith's; Martin's; Boat Trip through the Lakes; A Visit to the Southern Lakes; Big Tupper; Long and Raquette Lakes; John Brown Tract; Eckford Chain of Lakes; The Southern Adirondacks; Indian Pass; Elizabethtown; Lake Placid; Wilmington Pass and Whiteface Mountain.

CHAPTER IX.

ROUTES TO MONTREAL. — Via Lowell, Manchester, Concord; The Air Line Route; Lake Winnipiseogee; Down the Lake to Wolfboro'; Alton Bay; Plymouth; The Grand Trunk Route; Wells River; Newport; Lake Memphremagog; Trip down the Lake; Central Vermont Route; Mount Mansfield; St. Albans; Vermont Mineral Springs; From Rutland to Montreal; Plattsburg to Montreal.

CONTENTS.

CHAPTER X.

ROUTE TO NIAGARA FALLS.—Rochester to Niagara; The Village and its Hotels; Tour of the Islands; Biddle Stairs and Cave of the Winds; Prospect Park; Across the River; Suspension Bridge; Webster's Description; Across Lake Ontario; Down the St. Lawrence; The Thousand Islands; Ogdensburg; Down the Rapids.

CHAPTER XI.

MONTREAL.—Its Squares, Buildings, Churches, Institutions; Drive around the Mountain; Down the River to Quebec; City of Quebec and its History; The Suburbs; Montmorenci Falls; The Lower St. Lawrence and the Saguenay; Tadousac; Ha! Ha! Bay; other Trips from Quebec.

CHAPTER XII.

THE WHITE MOUNTAINS.—Old Man of the Mountain; The Flume; Twin Mountain and Crawford Houses; The Notch; Gate of the Notch; Climbing Mount Washington; North Conway; Kiarsarge Mountain.

THE PROLOGUE.

IN 1873 and 1874, Faxon's Handbook of Travel to Saratoga, the Adirondacks, Niagara and the Northern watering-places was received with the most flattering favor by the public generally and by tourists especially. So marked, indeed, was the favor accorded it, that the publisher would have been justified in re-issuing it without change, for an indefinite period, assured of abundant success. But the Spring of 1875 brought the completion of the Hoosac Tunnel, and other important routes of travel, which fact, with the desire to improve the book still further, and to place it at the head of all similar works, determined the publisher upon a complete revision of the work, including all changes up to the present time, and also introducing much additional information and many descriptions of scenes not found in the previous issues. This work, which extends to an entire re-writing of the book, has been carefully and conscientiously performed, great pains being taken to present the facts accurately, yet with due regard to interest, and it is believed that a thoroughly trustworthy guidebook, as well as an enjoyable volume for whiling away the tedium of travel is the result. The attention of the reader is especially called to several original features in this book which will be of value to the tourist.

CHAPTER I.

Steamboat Routes to Saratoga.

THE objective point of our journeyings for the present being Saratoga, we will first show how a very pleasant trip thither may be made principally by steamer, with more or less railroad worked in by way of connections, as may be desired. And in opening the consideration of this subject, several popular, estimable and every way first-class routes intrude themselves, so to speak, upon our notice, and we shall treat them, one at a time as they deserve. But first, whichever route be chosen, we will presume the tourist understands his business, has procured his tickets and laid out his course beforehand, and has not as some foolish virgins do, left all till arrival at the railway station, a few minutes before train-time. Thus in place of a nearly distracted and thoroughly irritated man, hurlling questions at the head of a busy and preoccupied ticket-seller or a hurrying and impatient porter, while his family (the traveller's, not the porter's) stand around in anxiety and despair, we shall have the pleasing spectacle of a complacent and amiable gentleman, leading a well ordered and joyous procession of his family into the right car, at the right time and bound in the right direction; with ample leisure for the purchase of the morning papers and the selection of seats on the shady side of the car, or a desirable section in the "drawing room" if preferred. This much premised, we will proceed to speak of

The Stonington Route to New York,

whence we will take the Hudson river boats to Albany, and the Rensselaer and Saratoga Railroad thence to the Springs. The agency in Boston of this line is at 228 Washington street, where courteous gentlemen will be happy to afford any necessary information and to sell tickets for any desirable excursion taking in this route. From Boston our course is over the Boston and Providence Railroad, one of the best-managed and most comfortable roads to ride over, in the country. The station is on Park Square, only a few steps from the Common and the Public Garden, and is one of the largest and finest railway depots in New England. The magnificent train house, a huge Gothic arch, accommodates a dozen trains at once, without confusion or inconvenience, while the waiting rooms, offices, etc., in the head-house fronting on the Square, are magnificent in their dimensions, furnishings and conveniences. A lofty clock-tower surmounts the head-house and adds beauty and lightness to the outward appearance of the structure. Taking seats on the right hand side of the cars if it be a hot and sunny day, we are whirled out of the station, across the Back Bay, stopping for one instant at the crossing of the Boston and Albany, and then a delightful ride begins through the beautiful western suburbs of Boston. The several neat and convenient stations in the newly annexed wards of the city are passed, and then come Hyde Park, Readville and Canton (14 miles from Boston, and the seat of one of Eliot's Christian-Indian settlements), where the railroad traverses a viaduct of solid granite 63 feet high and 600 feet long, and whence, seated in the cars, we overlook the roofs of the village. Here the Stoughton Branch, four miles long, leaves the main road and runs southeasterly to Stoughton forming a connection with the Old Colony. Blue Hill, an eminence over 600 feet high, commands a fine view of the city which we have just left, the ocean dotted with snowy sails and the whole surrounding country. Next come Sharon and Foxboro, noted for their fine hills, their manufacture of straw-goods and their fine farms. Mansfield, 22 miles from Boston, is the next place, and here we make our first stop, taking on board probably, a number of passengers who have come down from Fitchburg, Clinton, Lowell or South Framingham "and way stations" via the Mansfield and Framingham Railroad, which here crosses our track on its way to Taunton and New Bedford, and which will hereafter be described. The junction here of these two important lines has made Mansfield quite a busy, thriving railroad centre, though there is little else of interest about the place. Passing West Mansfield "on the fly" we are soon whizzing through Attleborough, on a high embankment, giving us a view of the busy village, with its factories lighted up, if it be a dark evening, and their hundreds of windows gleaming through the darkness like fire-flies. The principal industries — indeed almost the universal manufacture, is that of jewelry. The tradition used to be that Attleboro jewelry could be procured for fifty cents a peck, and that it was dear at that, but of late the manufacturers have copied costly gold jewelry with such accuracy and good taste, and have so thoroughly plated their goods that their appearance can be hardly distinguished from the original, and they will wear for a number of years. Dodgeville and Hebronville, manufacturing villages, are rapidly left behind, and then we pass through Pawtucket, 39 miles from Boston, the first town in Rhode Island, on a high embankment, with quite a sharp curve, and have a fine view of the many large and busy manufactories for which the place is famous. Here are the Pawtucket tack works, turning out some 300 or 400 millions annually, the Dunnell manufacturing company's thirty-six factories, where some 20 million yards of calico are yearly printed, the thread and spool factories, the steam fire engine works and many other industries. Pawtucket has its historical interest, also, being the spot where in 1676 Captain Pierce with 70 men were massacred by the Indians in the bloody "King Philip's war." From Pawtucket we ride only five minutes or so until we draw up in the fine and spacious railway station at Providence, 43½ miles from Boston.

Providence — Its History and Surroundings.

Probably every reader of this book is familiar with more or less of the history of this, the second city in size and wealth, of the Eastern States. "The State of Rhode Island and Providence Plantations," as the official documents read, has had more good natured fun made of it than any other State in the Union, except perhaps Delaware, on account of its tiny dimensions and its two capitals, being, since the rejection of the absurd custom by Connecticut, the only State which indulges in such an expensive and needless luxury. Yet if wealth, industry and prosperity count for anything, Rhode Island is a great State, and as for patriotism and national pride, the 1680 Rhode Island soldiers who fell in the war for the Union sufficiently answer. Providence "is the State" to a greater degree, probably, than is true of the capital of any other Commonwealth, not merely by virtue of containing some 70,000 of the 220,000 entire population of the State, but by virtue of its commerce, its manufactures, its educational institutions and the wealth and enterprise of its citizens. It is a beautiful city, most attractively located around the head of Narragansett Bay, which stretches southward to the ocean. The Providence river, which empties into the bay, expands into a cove, almost circular in form, and quite large in area, which lies to our right as we enter the station. This cove is surrounded by a broad walk, shaded by fine trees, amply lighted and protected by an iron railing, thus forming a favorite promenade in the Summer. The view across the cove in the evening, when all the twinkling lights are gleaming and their reflections flash from the water, is very pretty. The Seekonk river runs on the East side of the city, forming at its confluence with the Providence river a broad and commodious harbor. On the eastern side of the Seekonk, is East Providence, a pleasant suburb, through which the Providence, Warren and Bristol Railroad runs along the shore of the bay to Vue de L'Eau, and thence to Warren and Bristol. The city is very irregularly laid out, rivalling its ancient foe, Boston, in that respect, but its business streets and buildings are many of them very fine. The manufacture of steam engines and other machinery, silver ware, fire-arms, cigars, etc., foreign commerce and trade are the principal sources of the wealth of Providence. Alighting from the cars, we step out of the depot upon Exchange Place, where we see the fine soldiers' monument, from the design of Randolph Rogers, and which was erected by the State. The base is of blue Westerly granite and the superstructure of bronze. Four statues seven feet high represent the Infantry, Cavalry, Artillery and the Navy, and above them stands a figure of America, ten feet high, extending in one hand a wreath of immortelles for the fallen and in the other a sword and laurel wreath for the living. A short distance from the monument, towards the river, is the Custom House and Post-office, a massive granite building, and just across the river is the City Hall. The Arcade, a large, open hall, roofed with glass, on either side of which, on two stories, open retail stores of various descriptions, extends through from Weybosset to Westminster streets, in this immediate vicinity, and is one of the attractions usually shown to strangers. The Rhode Island Hospital, Brown University, the Athenæum, the Rhode Island Historical Society's Hall, the Dexter Asylum, the Butler Insane Hospital, the Friends' Boarding School, and the State Prison are the principal public institutions, and there are many fine churches of various denominations.

Providence was founded in June, 1636, by Roger Williams, who had been exiled from the Massachusetts Bay Colony for heretical religious views, he being a Baptist. On his coming hither, he was drifting down the Seekonk river, when at a rock near the foot of Power street, which is still shown to visitors, he was hailed by some Indians with the cry "What cheer?" He landed, and after a short confab with the savages, who were very friendly and amiable, he continued his course a few blocks further, under the India street bridge and around Fox street to the mouth of the Providence river, where he saw a good opening for a settlement, and accordingly settled. Since that day, Roger Williams and What Cheer have been the patron saints of Rhode Island, and nearly everything in Providence is named after one or the other. Roger Williams was a good man and he was the first to try the experiment of genuine and perfect religious liberty in this country. The consequence was that his little colony was soon filled up with all sorts of "damnable heretics," Quakers, Baptists, Catholics and those of no particular faith, yet they managed to prosper, branch out and increase, living at peace with each other and with the Indians. It will be observed that there were good Indians in those days. The Narragansetts, as Roger Williams found them, were a simple, amiable race, and what is most wonderful for Indians, industrious, supplying most of their dusky brethren of other tribes with wampum, pipes and pottery. The only good Indians nowadays, are dead Indians.

The suburbs of Providence are very inviting to the Summer tourist. By the Providence, Warren and Bristol railroad, one may visit several watering places along the east shore of Narraganset Bay, or may settle for a time at Warren, the former home of Massasoit, or at Bristol, near which is Mount Hope, the dwelling place of Metacomet or King Philip, son of the great sachem, and the bitterest foe of the whites in the long war which ended with his death. By steamers from Providence one can take passage almost hourly down the Bay to Vue de L'Eau, Smith's Palace, Silver Springs, Cedar Grove, Bullock's Point, Nayatt Point; Rocky Point, the most celebrated shore resort and clambake manufactory in New England; past Warwick, famous in old times as the seat of a colony of most remarkable heretics even for that age, and as the birthplace of General Nathaniel Greene, of Revolutionary fame; by Prudence, Hope, Patience and Despair Islands, then down between Rhode and Conanicut Islands, and into the harbor of Newport. From Providence, a line of steamers runs to New York, the Hartford, Providence and Fishkill Railroad runs west to Hartford and Waterbury, the Providence and Worcester northwest to Worcester, and the Stonington and Providence, by which we continue our route, and which forms part of the Shore Line, (all-rail) to New York, skirts along the west coast of Narraganset Bay and across a point of Washington County, across the line of Connecticut to Stonington at the easterly end of Long Island Sound.

Stonington and the Sound Voyage.

Leaving Providence we pass nearly south through the towns of Cranston (noted for its cotton-mills and for its Narragansett race-course), Warwick, previously described; Greenwich, the site of a Methodist seminary; Wickford, a sleepy, antique place, and Kingston (70 miles from Boston), the county seat, whence carriages convey passengers to Narragansett Pier, nine miles southeast, the youthful rival of Newport, with its cluster of hotels, its fine beach, its overlooking Heights, its morning bathing and afternoon croquet, and its drives to Narragansett Heights, the neighboring lakelets, Point Judith and other points of interest. This town of South

Kingstown is the largest in Rhode Island, covering an area of 76 square miles; it is noted as the birthplace of Commodore Perry and of Stuart the great painter, and for containing the great swamp in which was fought the decisive battle in the King Philip war. On a hill crowned with pines and cedars in the centre of this swamp are still to be seen the remains of the rude fort in which the desperate Narragansetts intrenched themselves and whence they were driven by the still more desperate colonists from Massachusetts and Connecticut.

Leaving behind Carolina, a manufacturing village, Richmond Switch and Niantic, we come to Westerly, which lies on both sides of the Pawcatuck river. Here is the dividing line between Connecticut and Rhode Island; accordingly one half the village is in one State and the rest in the other. Another curious feature of Westerly is that nearly all the inhabitants are Seventh Day Baptists, so that on Saturday the visitor will find the manufactories and stores closed and the church bells ringing. On Sunday, everything assumes its week-day aspect. There are extensive flannel and cotton mills here and the village has quite a picturesque appearance. There is one fine hotel here, the Dixon House, owned by and named after ex-Senator Dixon. From Westerly a little steamer runs twice daily down the river to Watch Hill, a favorite watering place on the Sound, or rather on the precipitous promontory which divides the Atlantic Ocean from the Sound. On the one side of this point the visitor can enjoy still-bathing; on the other surf-bathing, which in high winds is too high and strong for safety. Watch Hill is also accessible six times daily by boat from Stonington and once or twice daily by boats from Norwich and New London. There are seven fine hotels at Watch Hill, all on the summit of the Bluff, and a fine view of the Sound, Fisher's and Block Islands, and the town of Stonington on the mainland just across the sheltered bay. The collision by which the steamer Metis was sunk off Watch Hill, in August, 1872, will be remembered for many years by residents and visitors. The deck of the vessel with most of the rescued passengers washed ashore on the point, and so did most of the bodies of the drowned. The proprietors and guests of the hotels were indefatigable in their efforts for the comfort of the rescued.

Stonington, 92 miles from Boston, is the next station, but we do not see anything of the quaint and sleepy old town, for we are switched off a mile or two short of the station, and sent down to the landing alongside which lies the steamer Rhode Island, Stonington or Narragansett, with steam up, ready to convey us to New York. These steamers of the Stonington line (especially the Rhode Island, the newest of the Sound boats and one of the most elegant afloat) are famous the country over for their speed, safety, comfort and luxury, and the convenient hours at which they start and arrive have always made this a favorite route with the public.

The Rhode Island — A Model Steamer and an Enchanting Sail.

The Rhode Island may be taken as a model Sound steamer, her size, elegance and varied conveniences being united with speed and safety, thus making up all the desirable qualities of a steamboat. One novel and most agreeable feature is the location of the dining hall on the main deck, aft the space usually devoted to the Ladies Saloon. This dining hall is a spacious and luxurious apartment, fitted up tastily and looking out upon the water on both sides, thus ensuring a happy combination of light, fresh, pure air, and an everchanging scene upon the waters of the largest inland sea of America. The linen, silver, glass and service of this hall is rich, attractive, and, in beauty of finish, all that the most fastidious could hope or wish for. The dining room will seat 250 persons at one sitting. On warm afternoons, parties desiring it can dine off the spacious guards, a novelty in steamboat travel. There are 165 staterooms, each large and with lofty ceilings. Every room on the boat is lit with gas, and in each alcove is an electric bell, which communicates with the steward's department, which is a novelty on a Sound steamer. The rooms are richly furnished and fitted with every convenience. A Chickering grand piano graces the saloon which is sumptuously decorated, carpeted and frescoed, and is lighted by elegant bronze chandeliers. One of the noticeable features of this really magnificent steamer is the application of steam to her steering wheel, which reduces the chances of accident by collision to the merest minimum. A child can steer the Rhode Island, as far as strength is required. Under its present auspices the Stonington Steamboat Company has been in operation seven years, without missing a trip or losing a single life. It traverses what is called the inside route, thus avoiding the rough and uncertain passage around Point Judith, and its provisions for the comfort and pleasure of passengers are unsurpassed.

Embarking at about 9 o'clock we are soon steaming out into the Sound, with a view of the gleaming lights of the Watch Hill hotels on our left, and the blazing Fisher's Island light ahead of us. Soon we turn to the eastward and lay our course up the Sound, with the beautiful hills and green fields of Connecticut on our right, and the low, flat, monotonous shores of Long Island in the distance on our left. After a substantial supper in the saloon below, if it be a moonlight night, we shall find our chief enjoyment of the trip in sitting out upon the forward deck, watching the lights on shore, the pasing sails that gleam ghostly white in the moonbeams for an instant and flit by like morning vapors; the bold outlines of the eminences on the shore, or the islands along our course, while the gentlemen enjoy the acme of physical happiness in the whiffing of fragrant cigars, and the ladies, wrapped in fleecy nothings express their uncontrollable enthusiasm in positive declarations that "it's just too lovely for anything." Or, if the breeze be too strong, as Sound breezes even in Summer often are, we shall find in a seat upon the afterdeck, with the steamer's wake churned to frothy whiteness by the paddlewheels stretching behind us like a path of silver in the white moonbeams, a fascination that we would not forego, and here for hours we shall find delighted voyagers drinking in the beauty of the scene with placid contentment. But romance fades before the drowsy god, and we shall doubtless succumb to the desire for sleep sometime before midnight. Then we can retire to our cosy staterooms fitted with electric bells, running water, gas and other comforts of a first class hotel, and on a luxurious bed rest as sweetly as if at home, being wakened in the morning, if we desire, in season to enjoy the sail through the East River and Hell Gate; or if we prefer can slumber till the boat reaches her pier, No. 33 North River.

The Approach to New York.

The East River, so called, is simply the narrow strait by which the waters of the Sound communicate with New York Bay. The narrowest portion of this strait, filled with sunken ledges, projecting rocks and small islets, through which all the water is poured at every turn of the tide is known as Hell Gate, from its dangerous character in years past. Many vessels have been caught in its treacherous, boiling whirlpools and dashed upon its sunken rocks, to destruction. But the government engineers, by blasting out the submarine rocks have greatly changed the aspect of the place for the better and there is now little or no danger in navigating the river. Our entrance to the river is made where the Sound, suddenly narrowing, is almost cut off by the projection of Throgg's Neck, from the shore of Westchester County on the north, and the almost coincident projection of Willett's Point from Long Island on the south. Here the government has two strong fortifications commanding the passage of the river and the approach to the city from this direction. Soon we pass Flushing Bay, on the left, with the beautiful village of the same name at its head; Randall's Island, with its House of Refuge for young criminals; Ward's Island, with its Emigrant Hospital and Potter's Field; Hell Gate, with its swirling currents and rocky isles; Astoria and Ravenswood, pretty villages on the Long Island shore; Blackwell's Island, with its Lunatic Asylum, Workhouse, Almshouse, Penitentiary, Charity Hospital, Small-pox Hospital, and its neat little fortification, built by a crazy inmate named Maxey, who was impressed with the belief that this was the true point to defend the city. We now begin to realize our approach to the metropolis. The elegant villas and richly cultivated gardens on either side of the river begin to give place to foundries, ship-yards and other manufacturing establishments; on our left we pass in succession Hunter's Point, Greenpoint, Williamsburg (now part of Brooklyn), the Wallabout Bay, with the U. S. Navy Yard and the houses and spires of Brooklyn; on our right flit past one after the other Jones's Wood, the German festival garden, Bellevue Hospital, and then the solid squares of brick and mortar that go to make up the great city. Continuing down the East River, amid the swarm of ferryboats that dart out from the slips on either hand, by the forests of masts that line the wharves on both sides, we pass the huge and towering piers of the Brooklyn Suspension Bridge, turn to the right and are soon rounding the Battery. This small island on our left, covered with fortifications bristling with guns, and surmounted by a circular fort that looks as much like a cheese as anything, is Governor's Island. That round, odd looking structure on our right, with a conical roof that looks like a big gasholder, is Castle Garden, once a fort, later the fashionable concert hall and ball-room of the city, where Jenny Lind, Parodi, Sontag and other old-time prime-donne made their most notable successes; now the emigant depot of the city. It was formerly isolated from the main land, and accessible only by a bridge, but among the improvements carried on by the Tammany ring, with Boss Tweed as its centre, was the extension and beautifying of the Battery, by which Castle Garden was included within its limits, the whole territory enclosed by a splendid granite sea-wall, the surface graded, turfed and laid out in walks, trees planted, lights set, and the whole made a most attractive pleasure park for the densely crowded district in its vicinity. "Give the devil his due" is an old proverb, and Boss Tweed and the Tammany ring have received so much just denunciation that they ought to have credit for this genuine improvement to the city. Looking directly north from the Battery extends Broadway, the great artery of Manhattan Island. We are now entering the North or Hudson river, and passing between New York city on the right, and Jersey city and Hoboken on the left. On both sides are the piers and docks of steamship lines, foreign and domestic, and all along the wharves are the proofs of the immense commerce of the city. The piers are numbered in regular order, beginning at the Battery, and as our Pier is No. 33, it is only a short time before we are "warping in" and soon we are landed at the foot of Jay street, a few steps from West street. As this is not a cyclopædia nor a gazetteer, no description of New York city will be attempted; indeed to most people it will be unnecessary. We will simply pursue our journey Saratogaward, proceeding by steamer up the Hudson, as described in Chapter II.

The Old Colony Route to Saratoga.

Another favorite route from Boston to New York, *en route* for Saratoga, is that via Old Colony Railroad to Fall River or Newport, thence by the steamer Bristol or Providence to New York, and thence up the Hudson, as described in next chapter. If we decide on this route, our first move will be to visit the office of the line in the venerable old building at the head of State street, formerly the seat of the assembled wisdom of the colony, known as the Old State House. Having purchased tickets and secured check for a first class stateroom, we repair to the Old Colony depot, corner of Kneeland and South streets, a few minutes before half past four in the afternoon, and are soon comfortably seated in the cars. We have an opportunity to admire the fine station, with its lofty rotunda, its elegant and luxuriously furnished waiting rooms, its immense arched train house, its courteous ticket sellers and other depot officials, its complete system of designating trains and their starting time, so that no one could possibly go astray, and the numerous conveniences which all travelers must appreciate, but which are not often found in such perfection as here. If we choose we can fancy ourselves English lords or something else as we whirl over the smooth rails, by taking to ourselves a compartment in the English coaches which are run on the steamboat trains, but if we here have a patriotic horror of "blasted Britishers" and their ways, we shall find the ordinary cars sufficiently comfortable for the best Yankee citizen. We move out of the depot, and out of the city proper almost simultaneously as we cross Fort Point Channel to South Boston, on a pile drawbridge, but though out of the old town of Boston, we are not to be outside of the limits of the present city of Boston for some time. Crossing the South Bay on a causeway and pile bridge, we enter the old town of Dorchester, now the Sixteenth Ward, Boston. We successively pass Crescent Avenue, Savin Hill, Harrison Square and Neponset stations, all in the Dorchester District, and all on the shores of Dorchester Bay, across which fine views of the harbor and islands are obtained. Then we cross the end of Milton (Atlantic Station) and enter the famous town of Quincy, famous as the home of the Adams family and the birthplace of Quincy granite. The three stations in this town are Wollaston Heights, Quincy and Quincy Adams, respectively $6\frac{1}{2}$, 8 and $8\frac{1}{2}$ miles from Boston. The Quincy station is near the homestead of the Adams family. Then Braintree is passed, whence the South Shore division branches off to the eastward, passing through Weymouth, Hingham, Cohasset, Scituate and Mansfield to Duxbury, and thence to Plymouth by a short connecting branch, and we draw up for a moment at South Braintree, $11\frac{1}{2}$ miles from Boston. Here is a general junction, as three divisions of the road branch off here, one via the Abingtons, Hanson, Halifax, Plympton and Kingston to Plymouth, with a branch diverging from South Abington to Bridgewater on the second or Cape Cod Division. This division taking South Braintree as its point of departure, passes through Holbrook, East Stoughton, Brockton, Bridgewater, Middleboro, (whence two branches lead to the westward, one to Weir Junction, near Taunton, the other *via* Myricks, where the New Bedford Railroad is crossed, to Somerset Junction, forming a connection with the third or Fall River and Newport Division, yet to be described, through South Middleboro, Tremont (connecting with the Fairhaven Branch Railroad to New Bedford), Wareham, Cohasset Narrows (hence still another branch runs south via Falmouth to Woods' Hole, where the Martha's Vineyard and Nantucket steamer is taken), Sandwich, Barnstable, Yarmouth (where a little branch runs south to Hyannis Port, on the south side of the Cape, a watering place of some note), then along the sandy ridge known as Cape Cod, through Dennis, Harwich, Brewster, Orleans, Eastham, Wellfleet, and Truro, to the extreme curving tip of the Cape, the fishing town of Provincetown, 120 miles from Boston. The third division is the one with which we have to do, and we take the most westerly course of the three from South Braintree. We pass through Holbrook, 14¾ miles from Boston, a "shoe town," Stoughton, North Easton, Easton, and Raynham, in which latter place the first forge in America was set up by the Leonard brothers, in 1652, and soon enter the station at Taunton, 34 miles from Boston. This is a city of some 20,000 inhabitants, on the Taunton river, which furnishes the power for many manufactories, thus disproving the ancient libel that Taunton water was too weak to run down hill. Miss Elizabeth Pool, of Taunton in Somersetshire, England, founded and named this city in the early days of the colony, but it was only a pretty hamlet in 1810. Now, there are the immense Mason Locomotive Works, the Taunton Car Works, the various tack manufactories which turn out about the only kind of *tax* popular with the public, the Taunton Copper Works, several brick manufactories, foundries, cotton mills and an extensive Britannia-ware manufactory. The centre of the city is the Green, a neat square, with fine buildings fronting upon it. Near by are the buildings and grounds of the State Lunatic Asylum, a pleasant and popular Summer resort, the City Hall, Public Library and several fine stone churches. Taunton is a quiet and thrifty place, and much pleasanter to the visitor than its younger and more energetic sister, Fall River. Weir Junction, where the New Bedford Railroad crosses our track, Weir, North Dighton, Dighton (near which the famous Dighton Rock, with its supposed Icelandic inscription is found), and Somerset are successively passed, and we arrive at Fall River, 50 miles from Boston.

Fall River and its Factories.

Here is the great spindle city of the country, ranking even Lowell. The mills stand in rows, one above the other, along the steep banks of the river which falls 136 feet in half a mile, and so thickly are they studded along this magnificent water-power that they completely hide it from view. Many of the mills, however, are run by steam-power. Print cloths are the principal manufacture, though all kinds of cotton goods and some woolens, are made. Over $10,000,000 are invested in the Fall River mills, and they furnish employment for over 20,000 operatives. Most of the factories are massive granite structures and rank among the finest of their class. The sad disaster at the Granite Mill, No. 1, in the Autumn of 1874, by which 20 or more operatives were suffocated in the burning structure, or leaped from the windows to a cruel death on the pavements below, is fresh in everyone's mind. Fall River is solidly built along the shore of Mount Hope Bay, with Mount Hope itself looming up on the other shore. The boundary line of Rhode Island passes just south of the city; formerly it divided it, but Massachusetts ceded some land around Pawtucket to "Litle Rhody," and secured the whole of the "Border City" for herself. The Fall River, Warren and Providence Railroad runs hence, northwest, to Providence, 16 miles. At Fall River, our train runs down to an extensive wharf, alongside which, with steam up, lies the magnificent steamer Bristol or Providence, waiting to convey us to New York. These vessels, which are perfect counterparts of each other, have exhausted the praises of hosts of writers. For size, speed, beauty and luxury of appointments, they are among the finest steamers ever launched, and each Summer the crowds that patronize them attest their popularity. A fine military band on each trip performs selections on deck, and also dance music in the saloon, and the hours of the evening often wane into the "wee sma' hours ayant the twal," before the happy voyagers seek their sumptuous stateroom couches. From Fall River, our course is across Mount Hope Bay and into that of Narragansett, down which we steam for 20 miles, and round into the harbor of Newport. The sail down the Bay is most exhilarating and delightful, much more so than the ride by rail from Fall River, via Tiverton, Bristol Ferry, where the track crosses a narrow strait to the upper end of Rhode Island (the island, not the State), and then down to Newport, which occupies the southwestern portion of the island.

Newport and its Attractions.

This famous watering place, famous alike for its mild and equable climate, its magnificent ocean views and its refined and cultured society, can have but an imperfect mention here. A volume alone could do it justice. Indeed, many volumes have been devoted to the task and have only in part succeeded. Newport is one of the oldest of American summer resorts, and will always hold its preeminence, though since the war, the decline of Southern travel has tended to change the preponderance of the transient population from the great hotels to the magnificent villas or the cosy cottages that spring up like the work of enchanters all over the peninsula; from mere butterflies of fashion to refined and elegant summer residents. In the 16th century, Verrazani, a Florentine, visited this spot and wrote of its beauties, but even his landing is antedated by that of the Norsemen, if the testimony of the old stone tower in Touro Park may be credited. This wonderful ruin, which has set all the antiquarians by the ears for centuries, is a circular structure of stone, supported on round arches, and now covered with ivy and enclosed by an iron fence. It certainly does not bear out in construction, material or style of architecture the theory that a colonial governor built it for a wind-mill, in the 17th century, neither is there any record that such was the case, and the opinion generally accepted is that it was a watch-tower, built by the Norsemen who are supposed to have settled this section in the 11th century; the same who inscribed the Dighton rock. Its elevated locality, its workmanship and its style of architecture all tend to bear out this supposition. It is naturally one of the chief points of interest in Newport, and standing as it does, in front of the Atlantic House and near the centre of the city, is observed by all visitors. The old town, built around a fine harbor opening from Narragansett Bay, is a sleepy, antique-looking old burgh, with several buildings dating back before the Revolution, and a general air of musty tradition. The new town, on the elevated ground encircling the old part, and on the ocean shore forming the southern extremity of the island, is the fashionable Newport. Here are the magnificent, broad, hard, smooth and tree-bordered avenues, brilliant each afternoon with processions of stylish equipages; here are the splendid villas and the elegant cottages which the wealthy Summer residents from New York and Boston yearly occupy; here is the abode of the society which gives Newport its chiefest charm. It is not a place for a visit of a week or so, like Saratoga or Long Branch. One doesn't get into the ways of the *habitues* in that time, and one needs to visit Newport often, and stay a long time, to become familiar with its attractions and to enjoy its advantages.

The Sound Trip to New York.

But we have no time to revel in the delights even of Newport, and whether we have come hither by boat from Fall River, or being delayed have taken the later through express to Newport, we must be promptly on board our splendid craft which is impatiently snorting at her moorings. At last we are off, and steaming out of the harbor, between Goat Island (the seat of the United States naval torpedo station), and Fort Adams, on a point partially enclosing the harbor, we pass between Rhode and Conanicut Islands, into the Atlantic. Rounding Point Judith, famed in the past for rough weather and universally seasick passengers, but now, with immense steamers and the highest degree of comfort, little feared, we skirt along the coast of the mainland, with the state of Rhode Island on both sides of us, which seeming paradox is explained by the fact that while the state proper lies to the north, Block Island, noted for its codfishery, and belonging to the same gorgeous little State, lies to the southward some ten miles. Soon we pass Fisher's Island on the right and Long Island begins to overlap us on the far left. We pass the mouths of the Thames and the Connecticut, and lay our course straight up the Sound, arriving in the East River at early dawn, and at our pier, No. 28 North River, about sunrise. From New York up the Hudson, our route as is described in Chapter II.

The Norwich Line to Saratoga.

By the Norwich line of steamers, a very direct and easy route is afforded us from the Hub to the metropolis. For information, tickets or staterooms, we shall apply at the office of the line, 219 Washington Street, where we shall find every required courtesy and facility, and sometime before 6 P. M., if we propose to go via New York and New England Railroad, or before 5:30 P. M., if we go via Boston and Albany, we shall be on board the cars. By the latter course, we go direct to Worcester, thence over the Norwich and Worcester Division of the New York and New England, through Auburn, Oxford and Webster, Mass., and Thompson, Conn., to Putnam, Conn., 61 miles from Boston, where the train by the main line of the New York and New England joins us, and whence we proceed to Norwich and New London. By the New York and New England, we leave the station, foot of Summer Street, at 6 P. M., and trundling across the famous South Boston flats, on a causeway, we dash through South Boston in a deep trough underneath the streets and at the very roots of the houses, cross the South Bay and the line of the Old Colony, traverse the Dorchester District in a different direction and further inland than the Old Colony, passing the Stoughton Street, Bird Street, Mount Bowdoin, Dorchester and Mattapan stations, before we get outside the city limits. All these stations are in the midst of delightful rural scenery and have neat and attractive station houses and tasteful surroundings. Hyde Park, a station in the new and flourishing town of the same name, eight miles out, comes next, and then Readville, in the same town, where the line crosses that of the Boston and Providence. Then come Elmwood, Springvale and Ellis stations, all in the town of Dedham and all within thirteen miles of Boston. Dedham, the shire town of Norfolk county, is a quiet old borough with a considerable village in which stands the elegant court-house. In the township are several factories, power for which is afforded by "Mother Brook" so-called, though it is really not a brook, but a canal, and the oldest one on the continent. It was made in 1640, and its design was to increase the navigable facilities of Neponset River by diverting into it part of the waters of the Charles. It is three miles long, with 60 feet fall. Norwood, in the town of the same name, formerly South Dedham, Everett's, Winslow's and Tilton's stations are successively passed in the next four miles. All are thriving suburban villages, possessing much rural beauty, and all are largely inhabited by people doing business in Boston. Next comes Walpole, 19 miles from Boston, where the Mansfield and Framingham Division of the Boston, Clinton and Fitchburg Railroad crosses our track and affords through connections with Providence. New Bedford, Lowell and the north. Next comes thet own of Norfolk, formerly North Wrentham, noted for its straw factories, with the several stations of Campbell's, Norfolk and City Mills. Franklin is next, 27 miles from Boston, a town named for the immortal "Poor Richard," and by him presented with a valuable library. Wadsworth is the next station, and then Mill River Junction, 33 miles from Boston, a business centre of some importance from the fact of the Woonsocket Division,—which leaves Boston from the Boston & Albany Depot, and pursues a route through Brookline, Newton, Needham, Medfield, Medway and Bellingham—here crosses the main line, and affords connection with the great manufacturing village of Woonsocket, just over the Rhode Island line. Our next station is Blackstone, 36 miles from Boston, an important manufacturing village, just across the river (and the State line) from Woonsocket. Here the Providence and Worcester Railroad crosses our line, and affords yet another set of through connections. Our course now turns slightly to the northward, and we skim along through the southernmost towns of Worcester county, Blackstone, Uxbridge and Douglas, with the stations of Millville, Ironstone, East Douglas, and Douglas, all inside of 50 miles from the Hub.

Through Rhode Island and into Connecticut.

Then we trend southwest again, cut off a tiny corner of the northwestern town of Rhode Island, Burrillville, and cross into Connecticut, entering the northeastern town of that Commonwealth, Thompson. East Thompson is the station, and hence a branch 18 miles long runs northwest through Webster and Dudley to Southbridge, Mass., connecting at Webster with the Norwich and Worcester division. Our route next takes us through Thompson and Mechanicsville to Putnam, 59 miles from Boston. Here the Norwich and Worcester Division crosses the main line, and here is a large and commodious station, with all facilities for the comfort of passengers waiting for trains, and a good restaurant where an excellent dinner can be procured when desired. From Putnam the main road runs by Pomfret, Abington, Eliot's, Hampton, Goshen, and North Windham stations, among the most wild and romantic scenery of Tolland County, to Williamantic, the great railroad centre of Eastern Connecticut. Here the New London Northern Railroad from New London, Conn., to Grout's Corner, Mass., the Hartford, Providence and Fishkill Railroad from Providence to Waterbury, Conn., (which is designed to be one division of the New York and New England through line) and the New Haven, Middletown and Willimantic Railroad, which forms a direct connection with the New York, New Haven and Hartford Railroad, and New Haven, and thus completes the New York and Boston Air Line, (all rail.)

From Putnam to Norwich.

But our course does not take us to Willimantic, this time, as we turn to the left at Putnam, onto the track of the Norwich and Worcester Division, which, coming from Worcester, 26 miles north, through the towns of Auburn, Oxford and Webster, Mass., and Thompson, Conn., now continues to the south through the town of Killingly, noted for Indian legends and traditions of the old colonial days. It is a rich manufacturing town, having large woolen and cotton mills at Daysville and at Danielsonville, which latter is the principal village of the town, where two or three weekly papers are published, and where the arrival of a train brings together nearly the entire floating population at the station. Wauregan comes next, a small station, the village being some distance west, on the Quinebaug river, which furnishes power for the large Wauregan cotton mills, and near the pretty Quinebaug Pond, three miles long, connected with which is a legend that once in seven years, at midnight, a pillar of fire (known as the Narragansetts' fishing light), rises over the centre of the lake. The object of this apparition is not stated by the old settlers who claim to have seen it, but as they still live, apparently it is not a dangerous omen. We now pass into the town of Plainfield, with Central Village, a large cotton manufacturing hamlet, as our first stopping place, and then Plainfield Junction, 18 miles from Putnam, 78 from Boston, where we cross the Hartford, Providence and Fishkill Railroad. At Jewett City, a small station six miles further on, we cross the Quinebaug river, and soon after pass through a tunnel 300 feet long, cut from the solid rock. Greenville, eight miles further southwest, is a large manufacturing village on the Shetucket river, which we have crossed a mile or two back, near the point where the Quinebaug empties into it. From this point we follow the right bank of the Shetucket, two miles, and arrive in Norwich, the principal city of eastern Connecticut, and the shiretown of New London County.

Norwich, its Peculiarities and its Beauties.

Norwich is 94 miles from Boston, and is built on a lofty promontory between the Shetucket and Yantic rivers, which here unite to form the majestic Thames, and on the alluvial ground at the foot of the steep acclivity. In its location and configuration Norwich much resembles Quebec. No other place on this continent probably has a more commanding situation, or a more delightful overlook. In the construction of the town, also, Norwich is much like Quebec, barring the wall and the citadel of the latter. The commercial streets run along the Shetucket front, whose wharves are crowded with shipping; the residence streets are terraced one above another on the overhanging bluff, to which zig-zag lines of streets lead up like the "switchbacks" on a coal railroad. The promontory points to the south and from its summit, crowned with handsome villas, the view of the noble Thames which expands from its very foot and stretches way to the Sound, fifteen miles south, is charming and grand. Norwich, in consequence of its site at the head of the navigable waters of the Thames has an extensive commerce with coastwise ports and with the West Indies, while by its propinquity to the great cotton manufactories of Eastern Connecticut it enjoys a large trade in those articles. From the low ground on the river banks, near which the railway station stands, the view of the town is confused and meagre; one seems to be looking up into an overhanging mass of houses, with no regularity, but as one becomes acquainted with the place, and traverses its entire extent, it is seen to be a symmetrical and pleasing city of some 20,000 inhabitants, making considerable pretension to architectural beauty, and abounding in delightful walks and drives. By reason of its peculiar configuration, Norwich is one of the most oddly built cities, in its business portion, of any in the world, and has been made the subject of no end of jokes on this account. Norwich is the original town where the people of one street look from their kitchen doors down their neighbor's chimneys on the next street and see what they are cooking for breakfast; where a man steps out of his garret window into another man's back door yard, and where the cellar of one house drains on to the roof of the next below. Without joking, some business blocks which stand three stories high on Main street are six stories high on the next street, if they run through the block, as some do; while on the other, or up-hill side of Main street, a block three or four stories high is so built into the solid rock of the hill that you have to go up two flights of stairs before you see daylight at the rear, and the back yard is only accessible from the upper floor. As may be supposed, streets do not run up and down the hill, but around its side, gradually climbing upward, and instead of cross-streets, there are flights of stairs by which pedestrians get from one street to another, while vehicles have to take the long gradients and sharp angles by which the streets communicate at their extremities. Climbing above the business portion of the city, Washington street runs along the west side of the promontory, overlooking the Yantic and bordered on either side by magnificent lawns, on which, far back from the street, stand the stately mansions of the merchant princes. Broadway extends from the business center, nearly parallel with Washington street, but on the eastern side, and is less imposing in the matter of residences, and more like a city street, and both open upon the parade, (also known as Williams Park) a splendid field on the level plateau surmounting the bluffs, bordered with magnificent elms, and surrounded by elegant residences, among which is that of ex-Senator Lafayette S. Foster and the old home of General Williams of Revolutionary fame. On one side stands the noble building of the Free Academy which possesses a wide reputation, and broad streets diverge in various directions. Sachem street leads to the Falls of the Yantic, formerly a wild and romantic cataract, through a water worn gorge in the eternal rock, and over curiously grouped and massive boulders. The diversion of the water of the river by a dam above into an artificial channel, has left the rocky foundation of the Falls bare, except in seasons of freshets, yet the spot is always well-worthy a visit. But though the demands of business have thus marred the beauty of the Falls, they have widely increased their financial value. The artificial channel furnishes power for a score of extensive factories which manufacture paper, cottons, rubber goods and almost every thing else, and form the nucleus of the thriving Falls Village. Near the Falls is the old Indian cemetery, the burial-place for many ages of the Mohegan chiefs. Here is the monument to Uncas, the famous Mohegan Sachem, the steady friend of the whites, and with his army of trained warriors their most efficient ally against the Pequots, Narragansetts and King Phillip's confederation. Uncas was originally a Pequot chief, who in 1634 revolted against the Sachem Sassacus and joined the Mohegans. They made him Sachem and he brought the nation to the highest point among the tribes, and after ruling 50 years, died in 1683. In 1640 he sold the site of the present city to the colonists, for £70. Near Greenville, which is reached by horse-cars from Norwich, is the Sachem's Plain, where Uncas with 500 Mohegans defeated and killed Miantonomah, Sachem of the Narragansetts with 900 warriors. Five miles south of Norwich is Mohegan, the site of the fortress of Uncas, and here live the few half-breeds who represent the famous tribe.

From Norwich to New York.

From Norwich to the New York steamers, there are two routes, sometimes one and sometimes the other being used. By the first we cross the Shetucket near the upper end of the city, and proceed down the east bank of the Thames, through portions of the towns of Preston and Ledyard to Allyn's Point in the latter town, where are extensive wharves and depots for the eastern terminus of the Norwich and New York Transportation Company's fine steamers, City of Boston and City of New York. These boats rank among the finest and swiftest on the sound, and their course being entirely "inside" and sheltered by Long Island, the route is a favorite with many for whom seasickness has terrors. The other means of reaching the boats is via New London Northern Railroad to New London, thirteen miles, along the west bank of the Thames. The road follows the waterside closely, running at the foot of the steep and lofty banks most of the way, and the view from the windows at the left side of the train is of surpassing beauty. The stations are Thamesville, Waterford and New London, at which last place the train runs down upon a long covered wharf, alongside which lie the boats.

New London is a city of about 10,000 inhabitants, which has one of the finest harbors in the world, and in the days of the whale fishery was only second to New Bedford as an oil city. Of late it is one the sleepiest of decayed seaports, its wealth being mainly that accumulated by the whaling masters of a generation ago, safely invested. Its location, on a broad, gentle slope, is naturally very fine, and in the upper part of the city there are many fine residences and some handsome public buildings, but the business portion of the town is old-fashioned, out-of-repair, dirty and unattractive. The fishing and coasting interests and railroad connections are the principal sources of the prosperity (such as it is) of New London. It is a delightful resort in summer, despite the drawbacks mentioned, by reason of its natural advantages. The aristocratic Pequot House stands at the mouth of the Thames, some three miles south of the city, near the lighthouse, and other summer hotels are found in vicinity. Sailing, fishing and bathing are of the best and easily attainable, and steamers run daily to Watch Hill, Norwich, Sag Harbor and Hartford. Between the city and the mouth of the river, on a projecting rocky peninsula, stands Fort Trumbull, a massive granite fortress, manned by a small garrison, and on the east side of the river in the town of Groton, is a strong water battery. Near this point on the hights is the site of old Fort Griswold, captured in 1781 by a detachment of the force which under the renegade Benedict Arnold burned New London. The fort was defended valorously by 150 militiamen under Colonel Ledyard, who on surrendering his sword to the tory Captain Bloomfield, was run through the body with it, and the American prisoners were all at once massacred by their cowardyl captors. The story is familiar to every child who has studied the History of the United States. On the spot stands a fine granite shaft 127 feet high and 26 feet square at the base; a monument to the slaughtered militia. Above the city is the newest of our navy yards, secured by the Connecticutd elegation in Congress only after a severe struggle, and still in a rudimentary state. Besides the New London Northern, the Shore Line Division of the New York, New Haven and Hartfort Railroad has its eastern terminus here, forming a link in the Shore Line route from Boston to the metropolis. A ferry across the Thames to Groton conveys the cars back and forth.

Whichever route we take from Norwich, we will suppose ourselves safely on board the steamer and passing out of the Thames river by Fort Trumbull and the tall lighthouse, into the Sound. Moving to the right, we pursue the same general course as has been described for the other lines, and reach

New York early in the morning, where the Hudson river steamer is taken for Albany, as will be narrated.

The Shore Line Route to Saratoga.

The Boston and New York Shore Line is another favorite route from the Hub to the great city of Gotham, and by many travellers preferred to any other. We take the cars at the magnificent Boston and Providence station, and pursue the same course to Stonington as has been laid out in the description of the Stonington steamboat line. From Stonington we pass through Mystic, a small but thriving village, whose principal industry is ship-building, West Mystic, Noank, Pequonnock Switch and Groton, to New London, a distance of twelve miles. Near Mystic are Pequot Hill and Fort Hill, ancient strongholds of the Pequot Indians, who caused our colonial forefathers so much trouble. The former fortress was stormed in 1637 by a handful of settlers and a force of Mohegans under Uncas and Narragansetts under Miantonomah (this was before these sachems had fallen out) and the six hundred Pequot warriors were put to the sword. The view of the Thames river, the Sound and the neighboring country from Fort Hill is particularly fine. At Groton we get a fine view of the monument on the right, and soon the cars run down a gradual incline upon the immense ferry-boat which is in waiting to convey us across the river to New London. A large dining-hall is arranged on the upper deck, over the cars, and a capital hot dinner is served to passengers who desire during the transit. From New London we follow the shore of Long Island Sound, getting occasional fine land-and-water-scapes from the windows on the left. The shores are mostly low and sedgy, but there are some bold, rocky projections, and groves of fine trees. We pass through the towns of Waterford, East Lyme and Old Lyme (stations, Waterford, East Lyme, South Lyme, Blackhall, Lyme and Connecticut river) crossing at East Lyme the Niantic Bay, an arm of the Sound, on which is the beautiful village of Niantic, much frequented in summer, and provided with several good hotels and boarding houses. Boating and fishing are the principal attractions, and marvellous stories are told of the numbers of tautog and striped bass taken here.

An Historic Old Town.

We cross the Connecticut river by a covered truss bridge and enter the famous old town of Saybrook, at the station of the same name; the Connecticut Valley Railroad crosses our track, following the course of the Connecticut river from Hartford to its mouth, at Saybrook Point, near which is its terminal station of Fenwick Hall, a fine new summer hotel, with several handsome outlying cottages, and splendid views of river, Sound and shore, excellent bathing, fishing and sailing facilities, direct railroad communication, and the quiet elm shaded streets of the venerable hamlet of Old Saybrook near by. Saybrook Point was the site of an old fort built in 1635, and which was obliterated by the cutting through of the railroad. In 1636, Colonel Fenwick came from England and took command, bringing with him, his wife who died in 1648, and whose grave, marked by a rude stone, remained until 1872, when the ashes of Lady Fenwick were disinterred with considerable ceremony, conveyed to the cemetery and deposited beneath a monument. Saybrook was a place of note in colonial days, and the old fort did much service in repelling hostile fleets. Yale College was at first located in Saybrook, being chartered in 1701 and holding its first fifteen "commencements" here. In 1708 the celebrated Saybrook Platform was drawn up here, for the guidance of the college. We next pass through Westbrook, Clinton, Madison and Guilford, quiet farming towns, with watering places on the shores of the Sound, and come to Stony Creek, off which are the Thimble Islands, famous in tradition, and romantic in scenery.

The Romantic Thimble Islands.

Stony Creek has several small summer hotels, and a fleet of fine yachts lies in its little haven, for the accommodation of sailing parties or of visitors to the islands. These Thimble Islands are chiefly owned by New York and New Haven people, who have cottages of more or less simplicity upon them for their vacation residences. The group numbers 365, it is said, and all are rocky, bold prominences, rising out of deep water, and mostly covered with trees. The channels between the islands, by their depth of water and being sheltered from the winds by the islands, afford a safe haven to yachts and fishing vessels in storms, and many sail nightly make harbor among the isles. The pirate Captain Kidd frequently put in here, and Money island, the largest of the group, has been dug all over for the treasure he is fabled to have buried here. A small hotel has been built here by the owner, and most of the island laid out in building lots which have been bought up and occupied with small cottages. On Pot Island, the loftiest and best wooded of the group, is another summer hotel, where good living, pure air, the best of bathing and excellent views of the Sound can be had at a low price. The oysters which are taken from the waters of Stony Creek harbor and among the islands, are of wonderful size and flavor, and they form the staple of every meal if desired. Branford is the next station to Stony Creek, and is the point of departure for several popular summer resorts, Indian Neck, Pine Orchard, Branford Point and Double Beach, at each of which are one or more comfortable and moderate priced hotels. We next pass through East Haven, cross Saltonstall Lake, (whence the city of New Haven procures the most of its ice, and where a part of the Yale College class races are rowed) the Quinnipiack river, and passing through Fairhaven, (the great oyster mart of the section, and now one of the wards of New Haven) soon roll into the under-ground station in the heart of the Elm City.

New Haven and its Environs.

New Haven is the largest city in Connecticut, and up to 1873, when Hartford was very sensibly made the sole capital, divided the honors of the seat of government with that city. It has almost 60,000 inhabitants, is a thriving manufacturing and commercial town, and is famous for its magnificent elm trees, which line nearly every street. In the center of the city is the Green, a handsome public square, surrounded by elms and containing the old Center church, Trinity church and the North church, all ancient and venerable, and in their rear, on a gentle rise, the Old State House, now useless, a shabby old structure of brick and plaster, in the Doric style of architecture. Chapel street, the main business avenue, runs along one side of the Green, Elm street on the opposite, and they are intersected at right angles by Church street (on which stands the splendid City Hall), and College street completing the sides of the square, which is the size of four city blocks. Temple street, a broad thoroughfare, bordered by grand elms whose branches unite in a splendid gothic arch above it, traverses the Green midway between Church and College streets. Above College street are the grounds of Yale College and its fine and extensive buildings. This famous college, removed here in 1717 from Saybrook, is one of the chief seats of learning in this country, and its edifices number some fifteen large structures, besides several small society buildings and those of the Scientific, Law, Medical and Theological schools connected with the University. The Art Gallery is one of the finest and most extensive in the country, and the museums, reading rooms and other accessories are fully supplied and of a high order.

Manufactures, Suburbs, and Means of Transportation.

The manufactures of New Haven are so extensive and varied as to preclude particular mention in a work of this kind. The principal are firearms, of which the Winchester

Repeating Arms Company and the Whitney Rifle Company turn out immense quantities; church and parlor organs, pianos, carriages, agricultural implements, hardware, railroad cars, etc., etc. The city also enjoys an extensive West India and coastwise commerce, and is a very wealthy and intellectual city. The streets abound in handsome residences, nearly all of which, even in the heart of the town, have considerable grounds around them. On the avenues radiating from the business centre, are many splendid mansions, notably on Hillhouse avenue, at the head of which is "Sachem's Wood," a noble villa, surrounded by a park, the property of the Hillhouse family. There is a fine drive on the east side of the bay, at the head of which the city stands, by Forts Hale and Worcester, to the lighthouse, five miles from the city, at the entrance to the Sound. Savin Rock, four miles from the city, at the west side of the mouth of the bay is another resort. The road thither passes through the quiet old village of West Haven. The rock is a bold projection, near which is a fine beach, with a Summer hotel. East Rock and West Rock, some two miles inland from the city, are lofty, precipitous masses of trap rock, rising from the plain, from the summits of which grand views are obtained. Horse cars run to the base of each, and they are much visited. Near West Rock is Maltby Park, where is located the city water-works. There is a fine series of drive-ways in this park, which covers some 800 acres. Besides the Shore Line Division, the main line of the New York, New Haven and Hartford Railroad, the New Haven, Middletown and Willimantic, the New Haven and Northampton, or Canal Railroad, and the New Haven and Derby Railroad, all converge in the underground station, which will soon be a thing of the past, as the New York, New Haven and Hartford Railroad is now building near the basin at the head of the harbor, a magnificent depot, which will be, when completed, one of the finest in New England. Steamers run to New York twice daily, forming still another route to the metropolis, and small boats ply daily, in Summer, to the various resorts in the neighborhood.

From New Haven to New York.

From New Haven we travel by the main line of the New York, New Haven and Hartford railroad, which forms with the Boston and Albany the Express Line; with the New Haven, Middletown and Willimantic and New York and New England the Air Line; and with the roads by which we have come the Shore Line — all three direct all-rail express lines from New York to Boston. By whichever route he comes from Boston the traveller is transported from New Haven over the one trunk line, which, rolling out of the depot underneath the busy streets of the city, skirts the shore of New Haven bay in a southwesterly direction, passing through West Haven, Milford and Stratford, quiet country villages with broad, elm-shaded streets. We cross the Housatonic river, which divides the two last named towns, and are soon in Bridgeport, the fifth city in population of Connecticut, and one of the most extensive manufacturing centers of New England. Here are the sewing machine factories of Elias Howe, Jr., and Wheeler & Wilson, each employing several hundred hands and turning out many thousands of machines every year; a number of arms and ammunition manufactories, clock shops, and indeed manufactories of nearly all sorts of "Yankee notions." But Bridgeport's chief fame is that it is the birthplace of P. T. Barnum and Tom Thumb, and the former has a splendid mansion, Windermere, on the south side of the city, overlooking the Seaside Park, with its fine esplanade and its charming view of the Sound. Bridgeport has bi-daily steamers to New York, and the Naugatuck Railroad runs north-east to Waterbury, 62 miles, and the Housatonic north 110 miles to Pittsfield, Mass. Bridgeport is 18 miles from New Haven and 178 from Boston. We next pass through Fairfield, an ancient and quiet seaside village, which is rich in Indian and Revolutionary tradition, having been burned in 1779 by Tryon's Hessian Yagers, returning from a raid on New Haven. Southport, two miles further along, is chiefly famous for an Indian fight some 250 years ago, and has done nothing notable since. Westport is a young and vigorous village on the Naugatuck, and South Norwalk, 14 miles from Bridgeport and 192 from Boston, with its neighboring village of Norwalk, is the largest place between Bridgeport and New York. The chief manufactures are locks, knobs and hats, and the principal other trade, oyster raising and shipping. The magnificent million-dollar palace of the late Le Grand Lockwood stands near the village. At the drawbridge which we cross just before reaching the station, the fearful accident by which a train was precipitated into the river, took place, and caused the passage of the law now enforced, compelling the stopping of all trains before reaching a drawbridge. From South Norwalk, the Danbury and Norwalk Railroad runs north 24 miles to Danbury, the great hatting village of the country and the home of James M. Bailey, the Danbury News man; indeed of the Danbury News itself. Darien and Noroton come next, the latter being the site of Fitch's Home for Soldiers' Orphans, founded by Benjamin Fitch, with its fine library and art gallery. Stamford, 200 miles from Boston and 34 from New York, a very handsome village, is a great watering place with wealthy New Yorkers, and their cottages dot the hillsides on every hand. The New Canaan Railroad runs north-east to the adjoining town of the Biblical name, a distance of eight miles. Like old Canaan, it is "a land flowing with milk and honey," or would be if bee-farming were carried on, as it is a great cow country. From Stamford also, communication is had with New York by steamer. Cos Cob, near which Edwin Booth has a fine villa; Greenwich, where Israel Putnam performed his daring equestrian feat of riding his horse down stairs, and where are several fine churches and the famous Americus Club house of the Tammany ring, are next passed, and then we cross the frontier into New York State. Port Chester is the first station in the Empire State, a thriving village 27 miles from New York. Rye comes next, then Mamaroneck and New Rochelle, whence the Harlem River Branch diverges slightly from the main line, and runs to the Harlem river, through a number of the villages of West Chester County, which furnish residences for many New York business men. Our train passes rapidly by the stations of Pelhamville and Mount Vernon, whence we turn south-west and traverse the tracks of the Harlem Railroad, passing through the upper portion of New York City, with its shanty-crowned rocks, on which a large population of rag-pickers, pigs and goats sustain an unenviable existence; soon arrive at the numbered streets, and after a short ride draw up in the magnificent Grand Central station, 700 feet long, and covering three acres of ground. It extends from 42d to 45th streets, and half-way from Fourth avenue to Madison avenue. It is of brick, stone and iron, with several lofty domes, and miles of tracks inside its vast enclosure. Besides our own, the Harlem and the Hudson River trains enter and leave this depot, and from its vicinity various lines of horse cars and stages can be taken to any part of the city, or we can charter a hack direct to the pier of the Hudson river steamer which is to convey us to Albany.

The Boston and New York Express Line.

Another much frequented route from Boston is by the New York and Boston Express Line, over the Boston and Albany and the New York, New Haven and Hartford Railroads. Our route to Springfield is described under the head of the Boston and Albany route. From Springfield, we turn south. following the east bank of the Connecticut river, which spreads out from half a mile to a mile wide on our right, along the fertile alluvial meadows of Longmeadow, the staid and quiet village being on the high bluffs to our left. Below Longmeadow, near the Enfield Bridge station, and the old toll-bridge across the Connecticut, the river passes over a series of shallow rapids extending for several miles over the red sandstone ledges which form its bed. Near the upper end of these rapids, a wing-dam is thrown across the river, and diverts the water into a canal on the western side, which furnishes power for the various manufactories at Windsor Locks, some six miles below, where it empties into the river by a series of locks. Light draft, flat-bottomed steamers and scows by means of this canal make the voyage from Hartford to Springfield with coal, stone, etc. Formerly the traffic was very extensive, and embraced also the transportation of passengers by two or three daily lines of steamers, the passage being graphically portrayed by Dickens in his "American Notes." The passenger travel and most of the freighting has been monopolized by the railroad, of late years. Thompsonville, nine miles from Springfield, and 107 miles from Boston, is a busy manufacturing village on the east bank, in the town of Enfield. Here are the mills of the Hartford Carpet Company, the largest of the kind in the country, turning out nearly 2,000,000 yards annually. In this town of Enfield, the northernmost of Connecticut, are also the Hazard Powder Works, at Hazardville, and the famous Shaker community. At Warehouse Point, four miles south, we cross the Connecticut on a splendid iron bridge, built in England and set up here piecemeal on the piers of the old bridge, without interruption to travel, in 1866. We soon pass through the village of Windsor Locks, with its many paper, silk and iron mills, cross the Farmington river on a stone bridge of seven arches, pass through Windsor, a handsome old town of great fame in Colonial days, and now a great·tobacco-raising town, and in a short time enter the brown stone station at Hartford, 26 miles from Springfield, and 124 from Boston.

Hartford and Its Environs.

Hartford, the capital of Connecticut, is a beautiful city of 40,000 inhabitants, at the mouth of the Little River, so-called, and was settled by the Dutch in 1633, who built a fort at the mouth of the Little river, on a point still known as "Dutch

Point." Three years later, Thomas Hooker, a Puritan pastor, led his little flock through the wilderness from Newtown, and established himself here. A little log church was built near the present site of the old State House, and was the predecessor of a number of churches, of which the present lineal descendant is the old Center church near the same spot, in the burying ground of which rests Hooker's body under a stone table. The business center is on several streets parallel with the river, and intersected by streets running westward to the elevated ground beyond the railroad. The old State House stands in a small square at the junction of State and Main streets, on the eastern portion of which the United States government is building a post office. The new State House, a magnificent structure of granite and marble, is building on the gentle slope overlooking the City Park, the Little river and the railroad, on grounds formerly owned by Trinity College, which is removing to higher and more ample grounds, on the hills west of the city. The college is a wealthy Episcopalian institution, founded in 1823. The old buildings were of sandstone, with no particular architectural beauty; the new edifices will be worthy the college. The City Park is a splendid tract of 45 acres, bordered by the Little river (also called Park river) and is laid out in the highest style of landscape gardening, and adorned by fountains, statues and an elevated stone terrace. The splendid bronze statue of the late Bishop Brownell, on an eminence in front of the State House, the statue of Israel Putnam, and that to Dr. Welles, the discoverer of anæsthetics are its principal work of art.

Hartford's Public Buildings.

In strictly public buildings, Hartford has few of which to boast. The new jail in the northwest part of the city is one of the handsomest of them, and that is hardly attractive to the ordinary visitor. The Wadsworth Athenæum on Main street near the Little river is an old, castellated structure, with a gray mastic front, in which are the rooms of the Connecticut

Historical Society, including a museum of curiosities, chiefly dating back to Revolutionary and colonial times, relics of Indian wars, things that belonged to Israel Putnam, the bloody shirt in which Colonel Ledyard was slain, pieces of the Charter Oak, etc. Speaking of the Charter Oak, every other thing you see or hear of in Hartford is "Charter Oak" something or other — insurance companies, fire engines, eating houses, saloons, etc. Mark Twain says he has seen enough "genuine wood of the Charter Oak" to build a plank road to Salt Lake City, and he doesn't exaggerate much. The chair of the President of the Senate in the Old State House is made of the wood, beautifully carved, the museum above referred to has several pieces, and at least one family in town has a piano case made of it. And yet, the spot where the old tree stood is marked only by a round marble slab, a foot or so in diameter on Charter Oak avenue. The High School building on Farmington avenue occupies a sightly acclivity, overlooking the Park and the river, as well as the railroad and a good share of the city, while a little northward, the Blind Asylum on Asylum avenue, stands in a fine park, embowered in trees.

Magnificent Private Residences.

These avenues are largely appropriated by the residences of wealthy citizens. The Hon. Marshall Jewell, Postmaster General, owns a splendid mansion on Asylum avenue, and Mark Twain has the oddest of costly villas on Farmington avenue. The Retreat for the Insane is on Retreat avenue in the southwestern portion of the city, and to the eastward, on Wethersfield avenue is the magnificent estate of Mrs. Samuel Colt, widow of the great fire arms inventor. The property comprises a large tract of land, laid out as a park, with buildings for every conceivable use of a city mansion and suburban villa combined; conservatories, pheasantry, hot houses, graperies, and the like; while trees, fountains and statuary render the grounds among the finest in the country. Between this splendid estate and the river (from which it is divided and protected by a dyke), stretches the South Meadows, formerly pasturage, and bought very cheap by Colonel Colt as the site for his factories — now the property of the Colt Patent Fire Arms Company, of which General William B. Franklin is president.

Manufactories and other Business Interests.

On this vast tract, Colonel Colt laid out broad avenues and cross streets, surrounded it with a dyke to keep out the river in time of freshets, built cottages for his workmen, buildings for stores, a public hall, and in fact, laid the foundation of a complete village. His death, no doubt, somewhat retarded the progress of the new village, but it is still a neat, thriving and comfortable settlement, where most of the workmen in the arms factory and other industries connected therewith reside. On this territory, Mrs. Colt has built, as a memorial to her husband and deceased children, a splendid free church, of brown and Ohio stone, which for beauty of architecture and perfection of detail is probably unequalled by any church of its size in this country. The entire expenses are borne by this noble lady, who is as good as she is wealthy, and is universally beloved for her numberless acts of benevolence and Christian charity. Washington street, a broad, handsome drive, leading south from Capitol avenue, is bordered on either side by magnificent residences, fronted by green and velvety lawns. In winter it is the favorite sleighing ground of the city, and here the fast 'uns may be seen of an afternoon, if snow be on the ground. In its wealth, in proportion to population, Hartford stands at the head of the cities of the nation, as it does in respect to business buildings. The Phœnix Bank, the Connecticut Mutual Life, the Charter Oak Life and the Hartford Fire 'Insurance Companies' buildings are among the largest and most magnificent and can cope with any in the country The insurance business is pre-eminently Hartford's source of wealth, though trade, commerce and manufactures are largely responsible for the result. Besides the Colt works, the Pratt and Whitney Company, machinists, the Roper Arms Company, the Weed Sewing Machine Company, the Sharpe Arms Company and many other manufactories do a thriving business. Steamers run daily to New York, and others to the river ports, to Sag Harbor, New London, etc., and many sailing vessels come up the river to this port. Of railroads centering in Hartford, there are the New York, New Haven and Hartford, by which we have come and by which we pursue our course, the Hartford, Providence and Fishkill, from Providence to Waterbury, the Connecticut Valley from Hartford to Saybrook Point, at the mouth of the Connecticut river, and the Connecticut Western from Hartford to Millerton, where it connects with the Poughkeepsie and Eastern, and is designed to form a link in one of the through western lines. Hartford has a fine opera house, seating 1800, and another hall in which theatrical entertainments, concerts, etc., are given, several good hotels and all the characteristics of a live city. For its historic note, the reader is referred to any first-class history of the United States.

To New Haven and New York.

Having thus taken a random ramble about Hartford, we will return to the stone railway station and take passage on the next train south. For the first mile or so our route follows the curve of the Little river, which separates the track from the Pratt and Whitney, Roper and other machine works. We soon reach Parkville, a hamlet in the southwestern outskirts of the city, about the intersection of Park street with the railroad, near which is the Charter Oak Park, a fine enclosure with race track, designed for agricultural fairs and horse trots. Four miles further we come to Newington, where the Hartford, Providence and Fishkill Railroad, which has run on the same track with us since leaving Hartford, branches off to the right, and five miles more bring us to Berlin.

Hence, two branches diverge; the one to the right running to New Britain, a young but thriving city, noted for its manufactures of small hardware, tools, etc., and the other to the left, connecting with the city of Middletown, on the Connecticut river, 15 miles from Hartford, where also the Connecticut Valley and the New Haven, Middletown and Willimantic (Air Line) railroads cross. Middletown is a large, quiet and rather old fogy place, chiefly noted as the seat of Wesleyan University, the chief Methodist college of the north. The next stopping place is Meriden, a wide-awake, bustling city, the chief attractions of which are its numerous manufactories of silver-plated ware. Fire arms are also made here, and the State Reform School is pleasantly located on a slope in the outskirts. Three miles further, or 145 from Boston, comes Yalesville, a small manufacturing village; then Wallingford, noted for its britannia ware manufacture and its being the home of a branch of the celebrated Oneida Community of Free Lovers. The train next traverses a long stretch of white sandy plains, useful, no doubt in holding the world together, but as soil, too poor to raise mullein. The more of this land a man owns, the poorer he is. Geologists say, and there is every reason to believe, that this was once part of the sandy bed of New Haven harbor, and can trace the former shores of the bay in the rising ground bordering these plains. The road is terribly dusty here, and every body is glad to reach North Haven station, where we strike "solid ground" again. This is a great town for brick-making, and ships several millions yearly. Half a dozen miles more of travel amidst interesting scenery brings us to New Haven, whence we continue our journey to New York and thence to Saratoga.

CHAPTER II.

Up the Hudson River to Albany and Saratoga.

SARATOGA, however, being our present Mecca, and not New York, we will not delay in the metropolis, but seek the first conveyance to the Springs. Pier 39, foot of Vestry street, is but a few steps,—to be exact about it, half a dozen blocks,—and if we choose to do so, we can proceed direct thither and on board one of the splendid day boats Chauncey Vibbard and Daniel Drew, famed as floating palaces par excellence, for Albany. Or if we choose we may take a carriage for a short drive up town, or if desirable may snatch an hour or two for the transaction of business (this refers to the gnetlemen, of course) as the boat up the river does not start till 8.30. At that hour, accordingly, we shall be promptly on hand ,or if more convenient we may connect with the boat at the foot of 23d street, fifteen minutes later. We are soon comfortably ensconced somewhere on deck, so that our eyes can range the scene in every direction and get the full effect of the varied beauties of nature and art. As we head up the river, we leave behind us the crowded harbor and the bustling piers; to our left and rear is Jersey City, with its various depots for transatlantic steamships, its manufactures and its busy streets; directly abreast of us and adjoining Jersey City is Hoboken, the former picnic suburb of the metropolis, now a steamship and railway terminus, and a little beyond and on the same side is Weehawken. Straight ahead of us stretches the noble river, bearing on its bosom so great a proportion of the city's wealth-producing commerce; the vast fleets of canal boats, laden with coal from the Pennsylvania mines, or grain from the western fields; the steamers from Europe and the American coast ports; clam and oyster boats from the south and the lower bay; rafts of lumber from the north, and the pleasure palaces like that on which we are taking our passage. On our right is the great city, with its square miles of buildings, its labyrinths of streets and its forests of masts. As we proceed up the river we successively pass and recognize, if we be familiar with the metropolis, Manhattanville, with its Lunatic Asylum, Manhattan College, and the Sacred Heart Convent; Carmansville, with its Deaf and Dumb Institution, and group of fine villas; the Morris House, Washington's headquarters in 1776, Fort Washington, the highest point on the island, crowned with villas—all formerly suburban villages, but now connected portions of the great city. Between us and Mount Washington projects Jeffrey's Hook, the site of a redoubt in 1776, and on the west bank of the river, directly to our left, is Fort Lee, with its Revolutionary memories and its immense Palisades Hotel.

The Palisades.

The grand and wonderful Palisades, famous the world over, have begun to appear on our left, since passing Weehawken, and from Fort Lee for several miles north, they tower like a great wall above the river. These palisades are of the singular rock formation known as a "trap dyke," from 300 to 500 feet high, forming the west bank of the river. The lofty wall appears like a succession of vertical pillars, joined to each other, or the palisades of a fortification, whence its popular name is taken. Nearly opposite Fort Lee, on the East side, appears the mouth of Spuyten-Duyvil Creek, a tidal inlet, which with Harlem River forms a water communication between the Hudson and East Rivers, and isolates the island of Manhattan. The creek is crossed by several bridges forming the means of communication with the towns in Westchester County lately annexed to the metropolis. The legend goes that a Dutch trumpeter, Anthony by name, while on a journey in the old days of New Amsterdam, was impeded in his progress by this creek, then nameless. He swore that he would swim it, "en spuyt den duyvil," (in spite of the devil) and plunged in. But when half across, the veracious narrative goes, the devil angered at the free use of his name, came up in the form of a huge moss-bunker, or menhaden, seized Anthony and pulled him under, to rise no more. Above the creek we pass Riverdale, Mount St. Vincent, with its convent, and soon reach Yonkers, 17 miles from New York, a flourishing and beautiful village at the mouth of the Neperah river, with many suburban residences of New York merchants in its limits. Here was the ancient Philipse estate, the old Dutch manor house, built in 1682 and enlarged in 1745, being still in existence. Mary Philipse, the lovely heiress of this estate was sought in marriage by Washington long before he wooed the widow of Custis, and he never forgot her refusal. Hastings comes next, a busy town, and the port of shipment for the Westchester marble quarrjes. A little above is Dobbs's Ferry, an old village at the mouth of Wisquaqua Creek, and opposite is Piermont, on the line between New York and New Jersey. Hitherto, we have had the foreign country last named, on our left all the way, but now for the rest of our journey, we shall be in the United States and in New York, all the way. Inland from Piermont is the old Dutch hamlet of Tappan, noted chiefly as the place of Major Andre's trial and execution in 1780, after Benedict Arnold's unsuccessful attempt to deliver up West Point to the British.

A Region of Romance.

Here begins the Tappan Zee, a lake formed by the widening of the river, which is from two to five miles wide for a distance of ten miles. Near Irvington stands "Sunnyside," the old home of Washington Irving, to whose genius this whole region owes much of its charm, for he gathered up the quaint Dutch traditions that lingered about the scenes and localities, and interweaving them with the bright romances of his own brain, formed a chaplet which crowns the Hudson with immortal fame. The Tappan Zee and its neighborhood is the very center of this mythical and romantic region. In the legends of the early settlers the lovely lake is haunted by spectral ships of ancient Dutch mould, which came flying up in the teeth of the wind and tide, and never returned; by phantom whale-boats of the old water-guard, sunk by the British; and by the spectral skiff of Rambout Van Dam, destined to row between Kakiat and Spuyten Duyvil till the day of judgment. Even Sunnyside has its legendary interest. It was built over 200 years ago by Wolfert Acker, a Dutch councillor, who inscribed over the door, "Lust in Rust," (pleasure in quiet) and the English settlers with a droll humor nicknamed it "Wolfert's Roost." All around are beautiful villas, of New York grandees mostly, and the spot is charming to the highest degree. A short distance above is Tarrytown, the Dutch Terwe Dorp, immortalized in Irving's work, and near by is Sleepy Hollow, a quiet valley originally called Slaeperigh Haven, the scene of Irving's world-famous legend. Carl's Mill, the old Dutch church, built of bricks brought from Holland, the bridge over the Pocantico, where Ichabod Crane was overthrown by the Headless Horseman, the Philipse Castle, an old loop-holed mansion, built in 1683, as a point of defence for the tenantry of the Philipse manor, all are extant. Opposite Tarrytown is Nyack, and a short distance above is Sing Sing, a pleasant village on a sunny slope. Near the river bank, on grounds covering 130 acres, stands the famous State Prison, whose marble buildings were erected by the convicts who swarm like bees all over the enclosure. On the west bank is Verdritege Hook on Point-no-Point, a bold promontory on the top of which lies Rockland Lake, the ice-field whence the metropolis is chiefly supplied. Teller's or Croton Point projects from the right bank nearly two miles, as if to contest the passage of the river, and as we approach it we see the mouth of the Croton river, whence the water supply of New York City is conveyed 40 miles in a covered aqueduct. The dam is six miles up the river, and is 250 feet long, 40 feet high and 70 feet thick at the base, forming a lake five miles long, covering four hundred acres and holding 500,000,000 gallons of water. The aqueduct, of stone and brick, follows the course of the Hudson river to the great reservoirs in Central Park, and has a daily capacity of 60,000,000 gallons. The works cost $14,000,000, and include sixteen tunnels and twenty-five bridges, by which the conduit overcomes natural obstacles in its course.

The Highlands of the Hudson.

We round Croton Point, steering nearly towards the west bank, where the Highlands loom up grandly before us, and enter the beautiful Haverstraw Bay, a placid expanse of the river, named from the village of Haverstraw on the left. On Treason Hill, appropriately so called, stands the old stone mansion where Arnold and Andre met and arranged for the surrender of West Point. A short distance above, on the same side, is Stony Point, the scene of "Mad Anthony" Wayne's reckless, but successful assault in the Revolution, and opposite is Verplanck's Point, which he rendered untenable by the cannonade from Stony Point, after its capture. A few miles above, on the right bank, is Peekskill, at the mouth of a creek or "kill" from which the village takes its name,

Jan Peek, an early Dutch mariner having ascended hither and named the kill after himself. Here the river turns sharply to the left, and passes through "'The Race" so called, formed by the bluff promontory of Anthony's Nose on the North and the Dunderburg on the South. Between these the narrow channel is cut, the course of the river being very nearly from West to East for a mile or two. The scenery here is grand and majestic. Our steamer plows its way between the imposing mountains of the Highlands. The lofty Dunderberg (believed by the ancient Dutch to be the home of the storm-goblins — hence its name of "Thunder Mountain," as amusingly described by Irving) towers on our left, and hardly a stone's-throw on our right, Anthony's Nose (named, according to the same authority, from the bulbous and rubicund protuberance of Anthony Van Corlear, Governor Peter Stuyvesant's trumpeter,) rises 1128 feet above the water. Soon we turn to the right, pass Bracken Kill, Iowa Island, Poplopen Kill, and the remains of Forts Montgomery and Clinton, between which the Yankees in 1777 stretched a heavy chain and boom to stop the passage of the British fleet, but from whence they were driven by a flank movement of Sir Henry Clinton.

West Point and Above.

Buttermilk Falls are passed on the left, near which stands the famous Cozzens's Hotel, and soon we arrive at West Point, the nursery of Uncle Sam's incipient warriors, and the flirtation field of hosts of metropolitan damsels during the summer encampment. Here are the barracks for 250 cadets, the chapel, the hospital, the main academy building, the trophies of captured artillery, the ruins of old fort Putnam, and the Siege Battery near the water's edge. Across the river, we see Sugar Loaf, beneath whose shadow still stands the house in which Arnold made his headquarters; a little further along is Cold Spring, overlooked by Mount Taurus and Breakneck Hill. On the left side, a little beyond West Point, loom Crownest and Boterberg mountains, separated by the "Vale of Tempe," the scene of part of "The Culprit Fay." Near the northern foot of the last named mountain, lies the pretty village of Cornwall, and near by the former villa (Idlewild) of N. P. Willis.

Newburgh, Fishkill and Poughkeepsie.

But a short distance further, we see the busy streets, the white dwellings and the lofty spires of Newburgh, apparently climbing the steep bluffs on the west bank, while at the waterside stand blocks of huge warehouses, and acres of wharves and fleets of canal boats lie at the coal docks, loading with black diamonds brought direct from the mines by a branch of the Erie Railway. The "switchback," by which the loaded trains run by gravity down to the dock, and discharge directly into the canal boats, will be witnessed with interest by all our fellow passengers of a mechanical or material turn of mind; while the romantically inclined will find food for enthusiasm in the river and mountain view, and the lovers of history in the thought that here in Newburg, Washington had one of his numerous headquarters,—this one in a 'stone house over the heights, where he wintered in 1783—and perhaps will land and pay a pilgrimage thither, as to a second Mecca. Newburgh is an exceedingly sightly and handsome city, has 15,000 inhabitants, an immense coal and lumber trade, and is connected by ferry with Fishkill Landing, on the east shore, the western terminus of the New York and New England railroad, from Boston, via Hartford and Waterbury, if it shall ever be completed. The Duchess and Columbia Division, now runs to Millerton, where connection is made via Connecticut Western, with Hartford, but the link between Waterbury and this Western Division is still missing. A few miles north, and on the west bank, is a level rocky plateau, called by Hendrick Hudson, who witnessed there a midnight orgie of the Indians, "the Devil's Dance Chamber." Within the next few miles we pass Hampton, Marlborough and Milton, small and uninspiring villages on the left, and New Hamburg and Barnegat on the right, and next arrive at Poughkeepsie, 75 miles from New York, and the largest city between the metropolis and Albany. Poughkeepsie, on the east side, is a thriving and prosperous city of 20,000 inhabitants, and is the western terminus of another proposed line of railroad from Boston, to connect with an arm put forth from the West by the great Pennsylvania railroad. Vassar College, with its 400 young lady students, its splendid buildings and its unrivalled educational facilities, is about two miles from the city, and is an enduring monument to the benevolence of old Matthew Vassar. Besides this famous institution, there are the Poughkeepsie Female Academy, the Collegiate Institute, the Military Institute, ex-Mayor Eastman's National Business College, St. Peter's Academy, Cottage Hill Seminary, the Riverview Military School, and other educational establishments of a high grade, from which learning exudes, as it were, to benefit the whole country. There is also the State Hospital for the Insane, with its extensive grounds. The Poughkeepsie and Eastern Railroad runs east to Millerton, where it connects with the roads above named, and the Harlem from New York city, while the Hudson River Railroad, which has all along followed our course, on the right bank, passes through the city. The situation of Poughkeepsie is imposing and sightly. It is mainly built on an elevated plateau, far above the river, and its many fine edifices show off to good advantage as we approach and pass by.

The River Villages, and Hudson City.

Opposite Poughkeepsie is New Paltz, a landing connected with the city by a ferry; six miles above is the beautiful village of Hyde Park, near which point the river curves and narrows between high cliffs. This curve was appropriately named "Crooked Elbow" (Krom Elboge), by the ancient Dutch, and bears the name to this day. From this point, for a few miles, the scene is one of quiet beauty. Fertile meadows stretch on either hand, the river placidly expands, while the blue peaks of the distant Katskills form a pleasing background. Staatsburg, Rondout and Port Ewen, staid old villages, rich in history and tradition, are next passed Rondout is at the mouth of the creek of the same name, by which the Delaware and Hudson Canal makes its way to the Hudson. Kingston, two miles from the river, on the beautiful Esopus Creek, which has furnished the subject for many fine pictures, is a very old village, in which the first constitution of New York was framed, at a legislative session in 1777. It is now a place of some 7,000 inhabitants, and is reached by horse-cars from Rondout. Across the Hudson lies Rhinebeck, with a ferry connection, the main village being two miles inland. On a high bluff near by is an old fortress mansion, as manor houses were built in those days; the house of the Beekman family in the 17th century. Tarrytown, Tivoli, Saugerties, Clermont and Malden, river landings, are successively passed, each having something of interest in its history. Above Tarrytown is Annandale, the estate of John Bard, who has erected thereon St. Stephen's College, a fine stone Gothic building, for the education of young men for the Episcopal ministry; also the fine church of the Holy Innocents. Saugerties is at the mouth of Esopus Creek, marking the one hundredth mile from New York; Clermont is the ancient seat of the Livingston family, founded by the chancellor of that name, and Malden is the great shipping point of the North River flagging-stone. Passing Katskill Landing, the mountains of that name tower above it on the left bank, with the Mountain House plainly visible near one of the summits. From the landing stages convey tourists to the celebrated resort. On the way, the road leads through Sleepy Hollow, the scene of Rip Van Winkle's fabled 20 years' nap. On the east bank of the river, four miles above Katskill, is Hudson, the capital of Columbia county, a city of about 10,000 inhabitants. Here is the head of ship navigation on the Hudson, and hence the Hudson and Boston Railroad runs to Chatham, connecting there with the Boston and Albany. A few miles northeast are the Columbia Springs, often visited, and the views of the river, the Katskills, Helderbergs, Shawangunks, Highlands, and other mountains from Prospect Hill are very fine. Opposite Hudson is Athens, the shipping point of immense quantities of hay, ice, brick, etc. Here the New York Central Railroad has a great freight terminus. A short distance above is Coxsackie and then Stuyvesant Landing, New Baltimore, Schodack and Beeren Islands, Coeymans, Schodack, Castleton, Staats Island and Overslaugh are passed, and the steamer rounds to at the dock at Albany.

Albany, Its History and Its Attractions.

In 1614 the adventurous Dutch, who had even then sailed far up the Hudson and explored the magnificent country on its banks, deemed the site of the present city of Albany eminently fit for a settlement, and accordingly they settled. Nine years later they built Fort Orange, and called the little town Beaverwyck, owing to the numbers of beaver found here. In 1664 the British took the place and named it Albany in honor of James II, then crown prince, Duke of York and Albany. In 1686 a city charter was granted, and in 1798 it became the capital of the State. Albany, and indeed nearly all the county, and those of Rensselaer and Columbia, were embraced in the patent of 1150 square miles granted to Killian Van Rensselaer, by the Dutch East India Company in 1637 as Patroon of Rensselaerwyck, and here he and his descendants ruled in feudal state until the anti-rent troubles in 1846, when the state troops were obliged to interfere to put down the insurrectionary tenants, and since that time, the vestiges of the old system have disappeared, though the family still remains wealthy and famous. So much for history. The Albany of the present day is a thriving manufacturing and commercial city, doing an immense business by means of the Erie Canal, which here has its vast eastern terminal basin, with its breakwater 80 feet wide and 4,300 feet long, and by its railroad connections.

It is also the center of a great brewing interest, and Albany XXX ale is known the country over. The river is bridged for the passage of the Boston and Albany railroad, by a structure of stone and iron costing $1,150,000. Besides this railroad, which runs 201 miles east to Boston, the Hudson River 142 miles south to New York, the New York Central 298 miles west to Buffalo, the Rensselaer and Saratoga 94 miles northeast to Rutland, Vt., and the Albany and Susquehanna, 142 miles south-west, to Binghamton, where it connects with the Erie Railway, center here. The city has some 80,000 inhabitants and many fine public buildings. The most magnificent among these will be the new capitol, second only to its Federal namesake at Washington, if it is ever finished. It has been in progress many years, and has cost some $10,000,000. It is of light colored stone, in the Renaissance style, of which it is considered the finest example in the country. The water supply is drawn from Rensselaer Lake, five miles west, through a system of works costing $1,000,000. The marble State Hall, the City Hall, the Catholic Cathedral of the Immaculate Conception, St. Joseph's church, and St. Peter's (Episcopal), with its silver service given by Queen Anne to the Onondaga Indians, are all worthy of attention, as are the several educational institutions. But perhaps the most interesting building in the city is the old Van Rensselaer manor house, surrounded by its park, near Broadway, on the site of the original dwelling of the first Patroon of the name. This manor house is very ancient, and an interesting relic of the architecture of the feudal days of Albany. Here too is the old Schuyler mansion, built some two centuries ago, by the head of that distinguished and wealthy family.

A pleasant stopping-place in Albany, and one much frequented and enjoyed by tourists, is the famous hotel, Congress Hall, of which Mr. Adam Blake is the justly popular proprietor. Its location on the high land opposite the State House and the new Capitol; away from smoke and dust and noise of railroad trains and the business streets, makes it especially desirable to those fond of quiet, while comfortable, well furnished rooms, a sumptuous table and all the conveniences of a hotel leave little to be desired.

From Albany to Saratoga.

Continuing on our pilgrimage to the Springs however, we must drop the beauties and the traditions of Albany, and take the cars of the Rensselaer and Saratoga Railroad northward. Passing the Rural Cemetery, a little out of the city, we soon reach West Troy, the site of the Watervliet National Arsenal, with its hundred acres of enclosure, and its many substantial buildings. Across the river to our right, we see the city of Troy, with its fine buildings and its hosts of foundries. We soon reach Cohoes, a busy factory city of 16,000 inhabitants, at the great falls of the Mohawk river. Here is a costly dam built by the State, and by means of great hydraulic canals, water-power is derived equal to the task of manufacturing $10,000,000 worth of goods annually. Three miles above Cohoes, the Erie canal crosses the Mohawk in an aqueduct of stone with 27 arches, the whole structure being over 1100 feet ong. We cross the Mohawk river at Cohoes, and soon passing through Waterford, a manufacturing village, follow the left bank of the Hudson—a small stream above the confluence of the Mohawk, and shorn of all its grandeur— and traverse a long and fertile meadow between the river and the Champlain canal. Then we pass Mechanicsville, where are numerous thread factories, Round Lake, where the Methodists have a famous camp-meeting in "the season," and soon draw up at the station at Ballston Spa. This resort, though now less famous than its more northern rival, Saratoga, was in the past the great fashionable watering place of the country and still retains traditions of its former grandeur. It is now visited in the summer by many people, who desire a quiet and select, rather than a brilliant and showy company, and comfortable accommodations. There are several fine and famous springs here, among which the Sans Souci, in the grounds of the famous old hotel of the same name, is the most widely known. It is a spouting spring, and very rich in mineral virtues, containing 986½ grains of mineral matter to each gallon, 572 being chloride of sodium (called salt by some prosaic and vulgar people,) and 274 being bi-carbonates of lime and magnesia. The Artesian Lithia spring, bored in 1868, is considered a wonderful specific for rheumatism, gout, gravel and kindred diseases. It flows from a depth of 650 feet, and contains nearly 8 grains to the gallon of the bi-carbonate of lithia, and enough other mineral ingredients to aggregate 1,234¼ grains per gallon, making it probably the most strongly mineral water in the valley. The Ballston Springs are in the southern portion of the same valley with those of Saratoga, which seems to justify, in the wonderful efficacy, variety and quantity of its medicinal waters, the Indian superstition that here was the laboratory of the Great Spirit, where his children should come to be cured of their diseases. Ballston is the capital of Saratoga county, has several factories and a population of about 5000. From Ballston to Schenectady a branch runs southwest, the distance being 17 miles. Continuing on our northern course seven miles, we sweep by several grand hotels, forming the centre of a handsome and populous village, draw alongside an immense covered platform, and alight at Saratoga Springs.

Other Routes to Saratoga.

As already intimated there are several other routes which may be traversed on our way to Saratoga, but through lack of directness or the long time consumed, some special reason would be required for travelling them. One may go by Providence and thence by Hartford, Providence and Fishkill to Hartford, thence by steamer to New York; or by New York and Boston Air Line, via Putnam, Willimantic, Middletown and New Haven; or by steamer from Providence, or Saybrook, or New Haven, or even Bridgeport to New York; or by a dozen other routes or parts of routes; but, as already said, special reasons would be required to justify such a departure from the ordinary course of travellers. Some people, also, take the Hudson River Railroad from New York to Albany, but such people lose in great measure the enjoyment of the splendid scenery of the Hudson. But by whichever route we come, we will suppose ourselves to have landed safely under the long roof which covers the platforms and tracks at " the Springs," and to be receiving the congratulations and pressing invitations of the legion of delighted porters and drivers, who are unfeignedly glad to see us and anxious to take us to their respective hotels.

The Fitchburg, Rutland and Saratoga Line.

We will, however, describe one or two of the prominent all-rail routes thither. And first, we will premise that the tourist has purchased his tickets and obtained the necessary information as to routes, stopping-places, connections and time tables, both which desirable consummations can be reached by a call upon, or a letter to the office of the Fitchburg, Rutland and Saratoga Line, at No. 228 Washington street. The railway station on Causeway street, is the starting point for two routes, that *via* Fitchburg and Rutland coming first under our notice. Elegant and comfortable ordinary cars are furnished by this line, in which any one can ride as easily and with as little discomfort as on any road in the country. In addition, the famous Pullman cars are run on the through trains, in which those who are willing to pay for a little extra luxury and the attention of a special conductor can enjoy the acme of comfort in railway travel. As we leave the city and cross Charles River to Charlestown, the tall form of Bunker Hill monument towers above us on the right, and nearer by we see the grim walls of the State Prison. We barely skirt the edge of Charlestown, and then cross the Miller's river, pass through Somerville, Cambridge, Belmont and Waltham, at which latter place we see the immense works of the American Watch Company; Lincoln, in which town are the famous Walden woods and ponds, made famous by Thoreau's hermit life, and now the favorite scene of picnics; and soon we enter Concord, which was the scene of anything but concord one hundred years ago. On the 19th of April, 1775, history tells us, 800 British troops under Major Pitcairn, who had dispersed the patriots at Lexington the nigh before, were met at the North bridge across the Concord river by a little band of "embattled farmers," who "fired the shot heard round the world," and in so doing routed the proud hosts of the invader and sent them in disorder out of town. Here, upon the 19th of April, 1875, the centennial was celebrated with great pomp, a crowd of 20,000 people attending, the President and his Cabinet being present, an oration by George William Curtis, an address by Ralph Waldo Emerson and several other speeches being delivered, and the fine bronze monument of the Minute Man being dedicated on the old battle ground. But this is not a historical work of fiction; anyone who craves more history can consult the text books in the public schools. South Acton is the next station. It is chiefly noted as the point of departure of the Marlboro' Branch, 13 miles long. At Ayer (formerly Groton Junction) railroads from Worcester, Nashua, Lowell, Clinton and Peterboro', N. H., intersect, and here we are quite sure to receive accessions to our numbers from some or all of these places.

Fitchburg and its Environs.

Soon we reach Fitchburg, distant 50 miles (an hour and a half's ride) from Boston, and here we find another prosperous manufacturing town, an important railroad centre, and a delightful spot in summer to spend a few days or weeks. A good sized hill near the town bears the resounding title of Rollstone Mountain, and the brawling brook which courses through the village, supplying 25 water privileges and creating a necessity for several railroad bridges, is known as the Nashua river. From Fitchburg the Hoosac Tunnel Line branches off to the west, the Fitchburg and Worcester road runs nearly south to Worcester, and the Cheshire railroad proceeds (and we with and upon it) northwesterly to Keene, N. H. We pass through the towns of Westminster and Ashburnham, for the first few miles having fine views from the windows on the left of Wachusett Mountain, 2,018 feet high, in the northern part of Princeton, and passing in Ashburnham a number of clear, wooded ponds, which at this season are carpeted over with water-lily pads, and starred with their fragrant and snowy blossoms. Next we come to Winchendon, 68 miles from Boston, a large manufacturing town on Miller's river. This is the great hive of industry from which are produced myriads of wooden vessels, utensils and conveniences. It is no sign of illness in this town to turn a little pail, or a large one either, for hundreds of workmen are daily turning them, and "kicking the bucket" does not necessarily imply death. One of the largest of these factories, where everything wooden from a clothes-pin to a rocking-horse is turned out, is that of Captain E. Murdock, Jr., and it will be worth anyone's while to stop here and visit it. And they need not stop for that alone, for hence the Monadnock Branch makes off to Rindge, Jaffrey and Peterborough, N. H., and many people transfer themselves to this road for a trip to Mount Monadnock, of which more will be said hereafter. Others go by carriage from Winchendon, and, indeed, in all the neighboring country it is the custom to get up picnic parties to the top of Monadnock, for the benefit of sojourners from the cities.

Mount Monadnock.

This most celebrated peak of the vicinity is located in the town of Jaffrey, N. H., and is full in view from the car windows for several miles as we pass into the Granite State, either on the main line or the Monadnock Branch. It is a bold, rugged peak, 3450 feet high, nearly conical, and of great beauty when its harsh lines are softened by the distance. Near to, it shows wooded sides reaching nearly to the sum-

mit, though broken by perpendicular ledges of rock, and a
crest of solid jagged rock, bare and bleak as that of Mount
Washington itself. It is comparatively easy of ascent, and
one can leave Boston at 7.30 A. M., visit Jaffrey, ascend the
mountain, and reach home at 7 P. M., after a most delightful
and invigorating trip. From the summit of Monadnock, a
view, grand, beautiful and varied is spread out before the visitor. Southern New Hampshire and Northern Massachusetts
are at his foot, and though he does not see all the kingdoms
of the world and the glory of them, yet he gets a larger idea
of the greatness of this portion of New England than he can
from level ground. There are said to be 30 lakes embraced
within the range of vision, on one of the prettiest of which,
Contoocook, a small excursion steamer has been placed. The
Monadnock Mountain House, on the slope of the mountain, is
much frequented by visitors. Returning to the main line, we
pass through State Line, 71 miles from Boston, Fitzwilliam,
77 miles, Troy, 82 miles, and Marlboro, 86 miles, little towns
nestled among the hills, and favorite resorts for those who
were born under the last sign of the Zodiac. Then through
South Keene, 80 miles from Boston, a small station in the
southern part of Keene, and in a few minutes we roll into the
fine depot of Keene proper, 82 miles from our starting point.
Keene is one of the most charming towns in New Hampshire,
noted for manufactures, the power for which it derives from
the Ashuelot river and from Beaver Brook, the falls of which,
two miles north of the village, are an object of great interest
to the tourist. The location of Keene is beautiful, on a fine
meadow surrounded by hills, and traversed by the clear and
sparkling river. It is a town of 6000 inhabitants, has seven
churches and the county buildings of Cheshire county. The
Ashuelot Railroad runs hence southwest to South Vernon,
Vermont. The streets of Keene are broad, well shaded, and
the business centre, Central Square, has fine stores, in which
a large trade with the surrounding country is carried on.

From Keene we continue through the Westmorelands, the

first 100 miles from Boston, being oddly enough named East
Westmoreland and suggesting the inquiry as to whether there
may not be a North-East-by-South Westmoreland, then Westmoreland proper, 104 miles, Walpole 110 miles, Cold River
113 miles, the track following all the way the course of the
Connecticut River, and at the feet of a chain of magnificent
hills, the highest of which, Fall Mountain — a spur of Mount
Toby — towers 750 feet above our heads. Just beyond the
last named station we roll through a bridge which crosses the
Connecticut into Vermont, giving us fleeting glimpses of the
celebrated Bellows Falls, and in a few moments we are at
the Station of that name, an important railroad junction, and
one of the most flourishing manufacturing towns of the Green
Mountain State.

Bellows Falls, and Beyond.

The situation of this village is romantic in the extreme.
Looking to the eastward from the platform of the railway
station, Mount Kilburn, wooded with evergreen to its very
summit, towers like an emerald wall to the hight of 900 feet.
At its foot, and almost at ours, the river roars and foams. The
Connecticut is here compressed into a channel less than 50
feet wide, and the rush of waters through this narrow gorge
and over the huge rocks, which obstructs it, is magnificent,
especially during the Spring floods. Bellows Falls is a most
enjoyable place at which to spend a week, so numerous and
so varied are its objects of interest. To geologists, the
strange natural carvings of human faces in the rocks of the
vicinity will be well worth seeing; to the lover of fishing,
Warren's and Minard's ponds, Saxton's river and other places
afford fine sport, and if one enjoy witnessing the development
of the finny tribes, he may, at J. D. Bridgman's trout-breeding
establishment, ¾ of a mile north, gain all needed information;
to the invalid the Abenaquis Iron Springs offer their healing
waters, and to the lover of fine natural scenery there are
numerous pleasant drives, walks and climbs in the neighbor-

hood. Bellows Falls has been mentioned as a great railroad centre. Here, besides the Cheshire road, which forms a part of the great thoroughfare to Boston, the southern division of the Central Vermont Railroad comes in from South Vernon, where it connects with the Connecticut River Railroad, for Springfield, Mass. Here also, the Central Division of the same great corporation branches off to White River Junction, while we ourselves continue on northwest, over the Rutland Division of the same railroad. A ride of ten miles brings us to Bartonsville, 123 miles from Boston, where we begin the ascent of the Green Mountains, though we do not perceive any strong indications of our approach thereto before reaching Chester, 127 miles from Boston, whence we see from our post on the summit of a long green slope to the Williams river, a noble hill towering on our right. At Gassett's, 133 miles from Boston, we can take a stage, if we are so disposed, for Springfield, seven miles distant, and inspect the Black River Falls, which afford some wonderful illustrations of the action of water in wearing away rock.

Ludlow and the "Hog's Back."

Passing Cavendish and Proctorsville, which last place is noted for a quarry of splendid serpentine marble, much used for decorative purposes, we come to Ludlow, 141 miles from Boston, where we see the wonderful "Hog's Back." This euphonious designation is applied to a lofty ridge, whose formation has greatly puzzled geologists, rising abruptly from the green and fertile meadows. It is generally believed to have been an island in some primeval lake, before the breaking down of the eastern serpentine ridge drained off its waters and changed its bottom into a rich meadow. Over the crest of this ridge runs the railroad, ascending from Ludlow seven miles to Summit, the highest station on the road. At Ludlow we wait for the passage of the down train, and looking from our windows, up the grand sweep of the "Hog's Back," we see the train gliding like a serpent along its edge, and presently, with a roar like that of many waters, it emerges from a shallow cut and dashes up to the station at a speed which requires all the power of the brakes to check. From Summit to Ludlow no steam is used; the train is run by gravity alone; the speed is tremendous, and the sensation of riding over this portion of the road is exhilarating in the extreme. Ludlow is also famous for its magnetic iron ore and its fine beds of antique marble. Toiling up the "Hog's Back" we next reach Healdsville, 147 miles from Boston, a little station surmounting an ugly chasm, out of whose rocky sides several charming little cascades trickle. Summit comes next, one mile further, and here the dividing line between the eastern and western slopes is reached. Here steam is shut off, and we begin the descent to Rutland, 18 miles distant and 1,000 feet below us, our average descent being 55 feet to the mile. Mount Holly, East Wallingford and Cuttingsville are passed without comment, unless one should chance to notice from the windows Shrewsbury Peak, near the latter station, a commanding mountain 4,086 feet high.

Rutland and its Attractions.

Clarendon, 160 miles from Boston is the last stopping place before reaching Rutland, six miles further on, where we enter a large and handsome depot, and can, if we desire, get an excellent dinner at the restaurant in the station, or can patronize one of the hotels near-by, whose merits are loudly eulogized by a host of porters, whose friendliness and desire for the travelers' comfort are touching in the extreame. Many persons will desire to stop at Rutland for a time to rest from the fatigues of the journey, or to enjoy the fine scenery and the many attractions in the vicinity. For those fond of mountain climbing, Shrewsbury and Killington Peaks, lofty protuberances of the Green Mountain system, easily visited, will prove great attractions; Capitol Rock, on the north side of the latter, being a noted curiosity. For those whose blood is out of order, and who consequently "enjoy poor health," there are the Clarendon Springs near-by, whose waters, abounding in nitrogen gas in solution, sulphate and muriate of lime, sulphate of soda, sulphate of magnesia, a large amount of carbonic acids and perhaps other essentials to the prosperity of a well regulated drug-store,—are said to be equal as an alterative to those of the German Spa, which they much resemble. There are many fine drives about Rutland, to Sutherland Falls, to Killington, Shrewsbury and Pico Peaks, and to West Rutland where are some of the largest marble quarries in the world. Whole hills seem to be composed of solid marble, of snowy whiteness and fine texture, so fine, indeed that its value at the quarry is greater than that of Italian marble delivered at New York. A visit to these quarries is full of interest. Approaching, one sees first the great mill where gangs of saws are endlessly cutting the glittering stone into slabs of various thickenesses; the huge derricks, used for lifting the great masses of stone; then the piles of broken stone sloping away from the mouth of the quarry, like the piles of coal dust in front of a Pennsylvania coal shaft; then the gloomy opening to the mine itself. Some of these quarries have been worked for many years, and have completely hollowed out the interior of considerable hills, leaving only a shell of marble to support the soil and preserve the form of the outside. At first the work of quarrying was done by hand, but now steam is almost universally applied. Diamond drills, channelling machines and other like appliances are at work down deep in the bowels of the earth, the deafening noise of their action reverberating through the vaulted cavern; the gleam of light from the engine and the sooty smoke which constantly arises, and has in the lapse of years changed the snowy purity of the marble canopy overhead to inky blackness, make the whole scene to the unaccustomed visitor like a glimpse of Dante's Inferno. Rutland has several fine streets, handsome stores, hotels and churches, and the court house of Rutland County. The town has about 10,000 inhabitants and is very prosperous and thrifty. The railroads which centre here are the Rutland Division, Central Vermont, from Bellows Falls to Burlington and Essex Junction; the Harlem extension, running through Manchester and Bennington, Vermont, and Chatham Four Corners, New York, to New York City; and the Rensselaer and Saratoga, by which we take passage for the Springs,

The Road from Rutland to Saratoga.

Our first station is West Rutland, where the eye is attracted by the immense quantities of marble awaiting shipment, from the great quarries already described. This is also the nearest station to the Clarendon Springs, mentioned above, and here many visitors stop and take stages for the famous resort. Castleton, 10 miles from Rutland, 176 from Boston, a beautiful village of 1,000 inhabitants, is our next stopping place. This village is peculiarly favored by nature. It lies in the very lap of the Green mountains which rise abruptly on the east, while the rolling country to the west stretches away to the shores of Lake Champlain. It contains Lake Bomoseen, a clear and beautiful body of water nine miles long by three miles wide, dotted with islands and embosomed in lofty hills; its waters cold and limpid, the home of myriads of fine fish. Glen Lake, a tributary of Bomoseen, lies to the west, and Castleton river, a transparent, rapid stream, rising in the Green mountains, flows by its southern extremity on its way to join the Poultney river at Fairhaven, whence the combined stream makes its way to Lake Champlain over three falls aggregating 200 feet in hight. With such natural facilities, and the well known amiability and susceptibility to the blandishments of the angler of Vermont fish, it is no wonder that many gentlemen alight at Clarendon, whose principal baggage consists of rods, hooks, creels, lines and flies. But not alone do those of the male sex stop at Castleton. Fair and fascinating creatures, whom nothing less than a "Saratoga" or two can pacify in the way of baggage, are also dropping off here by every train.

They *say* they come to see Lake Bomoseen, to climb Bird, Herrick and Gilmore mountains and to visit the Falls, but there is reason to fear that they have designs upon the hardy fisherman whose prowess we sing; else why those clothes, why the scenes of gayety at the hotel, why the tales of flirtation that have become tradition in the neighborhoods? The road by which we came from Rutland lies through "The Gate," a narrow pass between the Bird and Gilmore mountains, of which we spoke above, and from Castleton we pass through Fairhaven, where we may stop, if so inclined, to visit the Falls of the Castleton river or continue to Whitehall, 26 miles from Rutland, 192 from Boston, a town of 6,000 inhabitants, at the southern extremity of Lake Champlain. During the French and Indian wars, and the Revolution, Whitehall, then called Skenesborough, was a place of much importance. The Champlain canal runs hence to Troy, connecting the lake with the Hudson river, and on our way to Saratoga the slowly-moving boats, with their tugging horses and efficient officers and crews, are in sight from the car windows a good share of the way. The Lake Champlain steamers have their southern terminus here, running hence to Ticonderoga, Burlington, Plattsburg, Rouse's Point and way stations.

An Historic and Legendary Legion.

The region upon which we are now entering is rich in historical and legendary lore. In the old days the possession of the Lake was deemed of such importance by all the powers that contended for the supremacy, that this territory was fought over again and again, and successively held by French, Indians, English and Americans. Every village, almost, bears the name of some fort, and has a crumbling ruin, or nearly obliterated earthwork, or at least a healthy tradition, to trot out in support of its claim. Fort Ann is the first of these villages, and it shows the remains of a redoubt erected in 1756, during the "Old French War," to command the head of boat navigation on Wood Creek. Next comes Fort Edward. Here was another stronghold, but the chief features of interest attaching to the place, are the legends of Jane McCrea and Major Israel Putnam. In 1777 during Burgoyne's invasion, Miss McCrea, the affianced bride of an American royalist in the invading army, was at the house of a friend near the fort. A party of the Indian allies attacked the house and butchered all the inmates save Miss McCrea. They took her with them towards the camp, but fearing pursuit, killed her also, threw her body into a spring and carried her scalp into the presence of Burgoyne, demanding the price of a traitor's head. The lady's lover was present and recognized the beautiful hair. His reason forsook him; he deserted the army, and wandering for a time he died by his own hand. The other legend is more cheerful. In the winter of 1757-8, Israel Putnam, then major, was quartered here. The barracks took fire near the magazine, where 300 barrels of powder were stored. Putnam mounted a ladder near the fire, ordered a line of men formed to the river and buckets of water passed as rapidly as possible to him, while he threw them on. Putnam stood there till the outer sheathing of the magazine was ablaze. Only a single thickness of plank intervened between him and death, but he wouldn't heed the entreaties to save himself, and just then the barracks fell in, the danger was averted, the fire was soon subdued, and peace and happiness reigned supreme. There is n't much of the fort left to corroborate the story, but it is doubtless true. It was just like "Old Put." At Fort Edward, we first touch the Hudson River, and here a branch leaves for Glen's Falls. Forty minutes more of riding, through a level and uninteresting country, and we arrive at Saratoga, 63 miles from Rutland, 229 from Boston. Our first intimation of approach to the village is the appearance of buildings on the left; then the most northerly of the springs is seen, the valley opens before us, and in a moment we whirl up to an immensely long covered platform at the station, and step out. We are greeted by hosts of hackmen and 'bus drivers, who, though strangers to us, hail us with a heartiness and a familiarity, which shows that they expected us and are delighted at our arrival.

The Boston and Albany Route.

By this route, we take the cars at the Boston and Albany depot, corner of Beach and Albany streets, and if we wish to go "through by daylight" and sleep in Saratoga the same night, we shall take the 8.30 A. M. train. Should we choose however, we shall be amply repaid for stoping by the way at several points, where most romantic and beautiful scenery is spread out. Securing seats on the right-hand or shady side of the elegant cars, or a cosily curtained section of the Wagner palace, we trundle out through deep cuttings alongside of or underneath busy streets, until the Back Bay district is reached, when a vast expanse of gravel stretches out like a gray-brown sea, over the flats once covered by sparkling waters, where 20 years ago or less little boys sailed boats, went in swimming and fished for mummychugs. A moment's halt is made at the "Know-nothing" crossing of the Providence railroad and then we are off again. Cottage Farms, Allston, where are the shops of the Boston and Albany railroad and the Beacon race track; Brighton, with all its great cattle yards and abattoir; "all the Newtons," lovely suburban villages, which give to strangers some of their pleasantest impressions of the vicinity of Boston; Auburndale, the site of the Laselle female seminary, a noted educational institution; Riverside, with its charming view of meadow and river and forest; Wellesley, with its beautiful Lake Wauban and the splendid Durant female seminary towering beyond; Natick, the house of Vice President Wilson, (whose modest white house is pointed out to visitors with as much pride as is the soldiers' monument in the little square near the depot), a town where shoes, hats and base-balls are manufactured, and in the early days of Massachusetts Bay Colony, the site of Eliot's Christian Indian community; Cochituate Lake on the right, whence an aqueduct of 20 miles in length conveys the water to the city of Boston, are passed, and soon we draw up for a short halt at South Framingham.

A Busy Railroad Centre.

At South Framingham, in addition to the vast through traffic of the Boston and Albany, is a very considerable railroad centre formed by the junction here of the several roads composing the system of lines under the management of the Boston, Clinton and Fitchburg. The main road runs northwest through Framingham Centre, a pretty rural village, noted chiefly as the site of the State Normal School, where every year a class of "sweet girl graduates with their golden hair" go forth to the easy conquest of the young men and the more difficult task of teaching the young idea how to shoot; through Southboro, Marlboro, Northboro and Berlin, pretty farming towns; through Clinton, a busy village at the junction of the Worcester and Nashua railroad; through Pratt's Junction, where the Fitchburg and Worcester is crossed, and Leominster, to a junction with the various northern lines at Fitchburg. The Mansfield and Framingham Division runs southeast 18 miles, through Sherborn, Medfield Junction, where it connects with the Woonsocket Division of the New York and New England road; Medfield, Walpole, where it crosses the main line of the New York and New England, South Walpole, Foxboro, and Mansfield, where it connects with the Boston and Providence, and the Taunton Branch, which the line has leased as a connection with the New Bedford Divison, over which passengers and freight are brought from the seaboard to the mountains. The Lowell and Framingham Division runs north 28 miles, through Sudbury, famed in colonial history and romance especially for its "Wayside Inn" of Longfellow's poem; through West Concord, Acton, Carlisle and Chelmsford, to the great spindle city, where another connection with the great northern lines is formed.

From South Framingham also, a branch of the Boston and Albany runs southward twelve miles, through Holliston to Milford, a busy village. Near South Framingham is Harmony Grove, famous for temperance, woman's, spiritualist and all sorts of "off color" mass meetings; also the Methodist camp ground. From South Framingham we pursue our course westward, following the Sudbury river some distance, and passing Ashland, Cordaville, Southville and Westboro, (at which latter place is the State Reform School, a water-cure establishment and the headquarters of the sleigh-building interest in this State,) Grafton, Millbury, (whence a branch track runs to the village proper 3 miles south, a busy manufacturing place,) and by a sharp turn to the right we come in view of the sparkling waters of Lake Quinsigamond, famed in years gone by as the course for the Yale-Harvard College regattas, and know that we are approaching Worcester. A few moment's ride, and we trundle into the splendid union depot, one of the finest, if not the finest railroad station in New England. It is of solid granite masonry throughout, "built to stand," and is 514 feet long by 256 feet wide, with a clock tower 200 feet high. Though built by the Boston and Albany railroad, it accommodates the trains of the Worcester and Nashua, Providence and Worcester, Norwich and Worcester, Boston, Barre and Gardner, and Fitchburg and Worcester railroads, and effectually supersedes the several isolated stations hitherto used by the various roads. It stands on Washington Square, a few rods north-west of the old "lower depot" of the Boston and Albany, and a quarter-mile south of the up-town or Foster street station.

Worcester and its Attractions.

Worcester is the second city in Massachusetts for wealth and population, containing over 50,000 inhabitants, thirty churches, many thriving manufactories and a number of academic institutions. It is built very near to the geographical center of the State, and accordingly is the place for holding most of the State Conventions of the political parties. Most towns are located by accident of settlement, by reason of natural advantages or some similar circumstance. Worcester was not thus placed, but was located and settled by order of the Colonial Legislature of Massachusetts Bay in 1669, as a half-way-house or halting place between Boston and the towns in the Connecticut Valley. Thirty families were located here and built a stockade against the "heathen," as Lo and his relatives were then facetiously termed, but "ye salvages" made it so warm for the settlers that in a few years the place was abandoned. In 1713 a new settlement was formed, a church built, which was also a citadel, and whither the male population carried arms and ammunition when attending services, — in fact, a genuine church-militant. Worcester was full of patriotism during the Revolution and sent a good sized regiment, the Fifteenth Mass., to the Continental army. A handsome monument to its colonel, Timothy Bigelow, stands on the Common. It was dedicated April 19, 1861, with a speech by Judge Thomas, the very day of the first bloodshed of the rebellion, when the Sixth Massachusetts marched through Baltimore. By a singular coincidence, the Fifteenth Massachusetts regiment, raised here, paraded and received its colors just 84 years after its namesake of the Revolution. Worcester sent several regiments and many fine officers to the Union army during the late war. The bravery of the fallen is perpetuated by a magnificent monument, dedicated last summer, consisting of a tall granite column, whose base is flanked by colossal bronze figures of soldiers, representing the several arms of the service, and its top is crowned by a globe sustaining a beautiful statue of victory. It has been mentioned that Worcester was not located by reason of natural advantages, yet had it been, a finer site could not have been chosen. Its manufactures have been developed by the power afforded by the Blackstone river which flows through it; its centrality and the conformation of its ground have made it a great railroad junction, while healthfulness and attractiveness are combined in its position among a group of romantic hills. Look which way you will, a graceful eminence rises before you, and generally it is crowned by the buildings of some college or academy. There are the Roman Catholic College of the Holy Cross, the Oread Seminary, for young ladies, the State Normal School, the Classical and English High School, the Free Industrial School, the Baptist Academy, the Highland Military School, and one or two others, all provided with fine buildings. Other objects of note in the way of buildings are the structure of the American Antiquarian Society on Lincoln square, with its library of 50,000 volumes, and its ancient portraits, its museum of old MSS. and curiosities etc.; the old Exchange tavern, where Washington and Lafayette have slept, and Mechanics' Hall, the largest in the State, where political conventions are held.

Westward from Worcester.

Continuing our westward journey, we soon arrive at West Brookfield, having passed Rochdale, Charlton, Spencer, East Brookfield and Brookfield, all smiling and fertile farming towns. West Brookfield is noted as the scene of a most determined resistance in 1675, by a little colony of Ipswich men, against the Nipmuck Indians. Huddled in a little garrison house, the brave colonists defended themselves for three days. Then the Indians, loading a cart with flax and straw, set fire to it, and pushed it up against the house. It had already began to blaze up, when a sudden shower extinguished the flames, and a gallant party of horsemen from Lancaster, 30 miles distant, galloped up and scattered the heathens like chaff. No wonder the old chroniclers considered that shower a miracle from Heaven. The various Brookfields are now noted for their shoe manufactories, and West Brookfield, where all trains stop, has a restaurant in the station, at which the best milk in New England can be procured. From West Brookfield, our course lies along the Chicopee river, which the track crosses a dozen times or so before reaching Palmer. Some of the glimpses of the river through the trees, dashing over brawling cascades, are very pretty. En-route to Palmer, we pass the Warren, West Warren and Brimfield stations, notable only for their manufactures and general thrift. Palmer is a stirring town, especially since it has become a great railroad centre. Here the New London Northern Railroad crosses our track, running from the Long Island Sound on the south to Grout's Corner on the north, where it connects with various northern and western lines. Hence also diverge the Ware River Railroad (leased to the Boston and Albany, and the Athol and Enfield.) Just across the Chicopee river, to the south, we see the huge, white building of the State Almshouse, in the town of Monson. Passing Wilbraham and Indian Orchard, with their factories, we descend a steep grade of about four miles, and enter Springfield.

Springfield and its Environs.

Springfield, 98 miles from Boston, 104 from Albany, and 135 from New York, forming the natural centre for the roads from those places, as well as from the several systems of lines to the north, via the Connecticut River Railroad, is located on the eastern bank of the Connecticut river, about two miles below the mouth of the Chicopee, and about the same above the mouth of the Agawam, which comes in from the West. It is a city of about 30,000 inhabitants, and is noted for its fine scenery, its railroad communications, its United States Armory and its manufactories of railroad cars (Wason Manufacturing Co.), of trunks and harness, of small arms, (Smith and Wesson Manufacturing Co.), etc. Springfield was settled in 1638, by a company under William Pynchon, and burned by the Indians in 1675, while the trainbands were absent defending Hadley. Only three block-houses, in which the inhabitants took refuge, remained standing, and one of these was extant at a very recent date. During the Revolution a shop for the repair of muskets, and a foundry for cannon were established here, and from this little beginning sprung the great United States Armory, at once the pride of residents and the wonder of visitors. From the depot, one may take a carriage, the street cars or omnibuses, or may walk down Main to State street, thence eastward half a mile, and up a rather steep ascent to the entrance, which is through a fine large gate, breaking the monotony of a splendid iron fence which encloses a beautiful natural park of 72 acres, on Arsenal Hill, a broad, nearly level plateau, on which stands the quadrangle of massive brick buildings, enclosing a large and handsome green. The works are at present on a peace footing, and employ only 500 to 700 men, but during the Rebellion the works ran night and day, some 3000 men found employment and about 800,000 stand of arms were manufactured. In the Arsenal, a large square building on the west side of Union Square, are stored 175,000 stand of arms, in solid squares, reaching from floor to ceiling of the lofty rooms. From the deck of the Arsenal Tower, a magnificent panorama spreads before us. On the north are Mounts Tom and Holyoke, twin sentinels at the gateway of the river, to the east are the lofty table lands of Willbraham and Ludlow, and to the west West Springfield's fertile farming lands, the sinuous Agawam river and the hills of Russell and Chester for a background. One mile south from the Armory proper, are the Water Shops, on the Mill river, where the heavy forgings and castings are made and where the gun barrels are tested. There are many lovely drives in and around Springfield. Crescent Hill, with the splendid residences of O. H. Greenleaf, J. G. Chase, Geo. B. Howard and many other wealthy citizens. Long Hill, Round Hill, where is the splendid villa of Dr. William G. Breck; Brightwood, with the summer residences of Dr. J. G. Holland (Timothy Titcomb) and George M. Atwater, and many other beautiful spots can be visited in a short circuit. The City Library on State street, is a handsome structure, exceedingly commodious and well arranged, from the designs of George Hathorne, the architect, of New York, who also built the splendid villas of Dr. Breck, O. H. Greenleaf and others mentioned above, and the fine buildings of the Springfield Institution for savings, corner of Main and State streets, the most prominent corner in the city. The new court house on Elm street, running through to West State and fronting on Court Square, is a massive granite structure costing $200,000. It has a tall clock tower, balconies, etc., the Italian style. The North and South Congregational churches (both new), the church of the Unity, the Roman Catholic Cathedral of St. Michael, the Memorial church at the base of Round Hill and the new State street Methodist church are the finest religious edifices. The new High School building on State street, opposite the City Library, is a large, fine and commodious structure of brick with granite trimmings, and a lofty tower. There are two fine hotels, the Massasoit, close by the depot, long famous for its table, and the Haynes House, corner of Main and Pynchon streets, down town, away from the noise and smoke of the railway, and everyway a first-class house. Its owner, Tilly Haynes, has done as much any one man for the growth and prosperity of Springfield, and is the owner of the neat and cosy opera house on the opposite corner. Hampden Park, the famous scene of the Springfield races, lies on the alluvial meadow near the river, north of the city, and one of the finest courses in the country, for a rowing regatta is that just below the city. Two or three miles north are the villages of Chicopee and Chicopee Falls, noted for their cotton manufactures, and the former as.well for the Ames Manufacturing Company's works, whence were turned out thousands of swords and hundreds of cannon and equipments during the rebellion, and where were cast the superb bronze doors of the Senate at Washington, the equestrian statue of Washington in the Boston Public Garden, and the statues of many Soldiers' monuments.

Up Through the Berkshire Hills.

Leaving Springfield the train crosses the Connecticut river on a splendid iron open bridge recently built, passes through the southern portion of West Springfield, along the left bank of the Agawam or Westfield river, through Westfield, (the Indian Woronoco), and famous for its manufactures of whips (a quarter of million yearly), and genuine Havana cigars (of Connecticut Valley seed leaf, — 10,000,000 to 12,000,000 a year). For its State Normal School with its 200 embryo school ma'ms, and its Soldiers' Monument, surmounted by a bronze statue in heroic size. Here the New Haven and Northampton Railroad crosses our line, and furnishes another connection between the mountains and the seaboard. From Westfield we ascend the valley of the river, with Mounts Tekoa and Pochassic towering above us on the right, pass the small stations of Russell and Huntington, and stop at Chester, a considerable center of communication and traffic with the surrounding mountain towns, and the location of a valuable emery mine. From Chester, the steep ascent of the mountains begin, and we climb for thirteen miles at the rate of 80 feet to the mile. Some long stretches have gradients of 82 feet. Becket, Washington, the highest point on the road, Hinsdale and Dalton, hill towns, famous for their rocks, romantic scenery and exhilarating air, are passed, and we arrive at the elegant station in Pittsfield.

Pittsfield and its Beauties.

Pittsfield, the capitol of Berkshire county, is one of the youngest cities in the State, with about 12,000 inhabitants, many noted manufactories and a great variety of beauties, natural and artificial. It is 53 miles west of Springfield, and 151 from Boston. Among the former, must preeminently rank its ladies, the fame of whose beauty extends throughout the Commonwealth. No susceptible bachelor can risk a visit to Pittsfield, especially to the vicinity of Maplewood Institute, if he expects to get home heart-whole. In the same class (natural beauties), must, of course, be placed the mountains and lakes, which environ Pittsfield on either hand. The town is built on a lofty plateau, some 1200 feet above the sea level, and the Hoosac mountains to the east, and the Taconies to the west, completely encircle it. Lake Ashley, whence comes the abundant and crystal water supply, Lake Onota, Lake Pontoosuc, Berry Pond, West Pond, Melville Lake or Lilly Bowl, Silver Lake, Sylvan Lake, and several nameless lochs are near the city, and are frequently visited. The Wahconah Falls, Lulu Cascade, South Mountain (whence a magnificent view is gained), and other hills and valleys are worthy of visits. New Lebanon Springs, a popular watering place, lies some 15 miles west, and on the way thither is the famous Shaker Village. Among the artificial beauties of Pittsfield the fine Soldiers' Monument, in the centre of the old green, attracts attention first. It bears a magnificent bronze statue by Launt Thompson, of a color bearer, standing on a massive pedestal. The dedication, September 24, 1872, called together the largest crowd ever gathered in Berkshire county, and was the occasion of a grand celebration. The elegant white marble Court House, the Berkshire Athenæum Building, the fine white marble Cathedral of St. Joseph, the Berkshire Life Insurance Company's Building, and the Maplewood Institute (surrounded by a magnificent grove), are among the finest public buildings. At Pittsfield, the Housatonic Railroad runing south, through Lenox, Lee, Stockbridge and Great Barrington, all famous for their magnificent scenery, enters Connecticut, and finally reaches Long Island Sound at Bridgeport. The North Adams branch runs north from Pittsfield, through Lanesboro, Cheshire and South Adams, and forms a connection with the Hoosac Tunnel line.

Across the Line and so to Albany.

Passing to the southwest a few miles, we reach the State line and cross the imaginary barrier into New York State.

From this point we traverse Canaan, which, though not "a land flowing with milk and honey" exactly, is a romantic and mountainous town enough, and soon arrive at Chatham. This place is particularly notable as the crossing point of the Harlem Extension Railroad, from New York City to Bennington and Rutland, Vt., and as the point whence the Hudson and Boston Railroad (leased by the Boston and Albany) branches off in a southwesterly curve to Hudson, a thriving city on the noble river of that name. From Chatham our course is a little west of north, through Kinderhook, Schodack and Clinton, the scenery growing less wild all the way to Greenbush, or East Albany, on the Hudson, which we cross on a splendid open bridge, and are landed in Albany. This city and the route hence to Saratoga have already been described.

CHAPTER III.

The Hoosac Tunnel Route to Saratoga.

HAVING left Boston from the Fitchburg depot precisely as described for the Rutland route, our course follows that till we reach South Ashburnham, 61 miles from Boston, where a "Y" sets us off in a more westerly direction, and we are soon speeding over the Vermont and Massachusetts Division. From Wachusett, stages run to Princeton, a few miles south, where is a popular Summer resort, and whence Wachusett Mountain, already mentioned, is ascended. Gardner and Templeton, hill towns, noted for their manufactories of wooden ware, are passed within ten miles distance, and we enter the fertile and picturesque valley of Miller's River. The railroad is elevated on an embankment, giving fine glimpses of Monadnock, the intermediate hills and the river at our feet. Baldwinville, Royalston, Athol, Orange, Wendell and Erving are passed in the next 20 miles. All are quiet little towns, noted for their rural beauty and for occasional bits of wild and romantic scenery, of which we gain only too brief views from the car windows as we fly past. At Grout's Corner, 98 miles from Boston, the New London Northern Railroad, from Long Island Sound, up through Willimantic, Conn., Palmer and Amherst, Mass., crosses our route, and continues northward through South Vernon to Brattleboro, Vt. The road follows the Connecticut river eleven miles to South Vernon, a small and unimportant village, and thence ten miles further to Brattleboro, a thriving manufacturing town, noted as the birthplace of Colonel James Fisk, Jr., and the place of his burial. A splendid monument to his memory, the work of Larkin G. Mead, the sculptor, was dedicated with imposing ceremonies on the 30th May, (Memorial Day) 1874, the Ninth Regiment of New York National Guard, which Colonel Fisk commanded, participating. Here a pleasant stopping place will be found at the Brattleboro House, owned by Jacob Estey, Esq., a prominent citizen, and ably managed by Mr. H. A. Kilburn, whose long experience in hotel keeping, insures the comfort of his guests. The house has been remodeled in excellent style and all the comforts are to be found here. Six miles from Brattleboro are the famous Guilford Springs. From Grout's Corner the route leaves the Miller's River valley, and passing through the town of Montague, crosses the Connecticut river on an open bridge, which affords a fine view in either direction of the beautiful stream and the romantic scenery of its banks. Just below this point, the Deerfield river empties into the Connecticut, from the west; we cross it and proceed up its valley to Greenfield, 106 miles from Boston, the county seat of Franklin county.

Greenfield and its Environs.

Greenfield is a beautiful village, rural in its appearance and charming in its location, though in its industries and its population it is a thrifty and prosperous town. It lies spread out upon the fertile meadows which border the Connecticut, the Deerfield and the Green rivers. The latter gives its name and furnishes the power to the Russell cutlery works, which employ some 600 men and turn out immense quantities of goods yearly. Here are also woolen mills, tool factories, etc., among the industries of the place. The square in the centre of the village is one of the prettiest examples of the rural New England "plaza;" its soldiers' monument in the middle being fronted and overlooked by the Town Hall, the Court House, and a fine stone church, while the sides of the square are filled in with handsome residences, stores, etc. There are, perhaps, as many objects of interest and places of favorite resort in the immediate vicinity of Greenfield as near any other place in New England. Not only have the numerous rivers which course down from the hills produced many objects of wild and romantic beauty, but tradition and history throw their charm over the region, and take us back to the early days when the settlers took their lives in their hands and only held on to them by the bravery with which they fought the heathen until they had driven them out. Near this place the Connecticut makes a descent of 36 feet, forming a waterfall, which Dr. Hitchcock pronounces the most interesting in the State. The Turner's Falls Company has here built a huge curved dam, and established manufactories which are expected to some day rival Lowell and Lawrence. Deerfield, a quiet farming town, lies five miles south of Greenfield. Coleraine, Leyden and Shelburne are towns in the immediate vicinity of which each has a gorge to boast; a dark, deep, narrow chasm, cut by the waters of the rivers which course through them, adorned with cascades and set in a framework of lofty hills. Arthur's Seat and the Poet's Seat are high and romantic hills overlooking the neighboring villages, the rich intervales, and the meandering streams. Leaving Greenfield, the road makes a detour to the south to avoid the disagreeable alternative of climbing over Arthur's Seat, and then, following the course of the Deerfield river, enters the dark and gloomy but majestic Deerfield Gorge, through which the river finds its way between Shelburne and Conway. So narrow is this defile that, before the construction of the railway, enthusiastic naturalists and lovers of the romantic could hardly pick their way on foot through it. At Shelburne Falls, 129 miles from Boston, the river leaps downward over a succession of giant stairs, which look as if hewn in the solid rock, finding a new level 150 feet below. Here is a large cutlery establishment. Passing through Buckland, Charlemont and Zoar, small hill-towns, noted for their bold and rugged peaks, their romantic glens, and the flavor of tradition dating back to the old Indian wars, we reach Hoosac Tunnel Station, 136 miles from Boston, and get our first view of the world-famous mountain and the yawning entrance to the " great bore."

The Hoosac Tunnel and its History.

The situation of the mountain (which is not only the great barrier between east and west, but, as it were, the centre of attraction to visitors among the northern Berkshire hills) may, at the outset, need some explanation. The Hoosac Mountain has two crests with an intermediate valley; the Deerfield river washes the eastern base and the Hoosac the western. Most tunnels are built on an ascending grade, as some descent is necessary to carry off the water; but these two rivers being at precisely the same height above tide-water, the mountain had to be entered at the same point at both ends, so that the only way to secure drainage was to have a summit at the centre, from which the grade descends about 26 feet in the mile to either portal. The crest overlooking the Deerfield valley is about 1,450 feet above the river bed; the Hoosac peak is 1,750 feet, and the lowest depression between these tops is some 800 feet above the grade.

Description of the Tunnel.

According to the terms of the last contract, which was agreed upon in 1869, and under which the tunnel was completed, one railroad track was to be laid through its length, and all unnecessary material removed by the first day of March, 1874. The tunnel was not fully in readiness for use at the time specified in the contract; but the contractors who undertook the work so faithfully performed their duty that the State did not insist upon too rigid a fulfilment of the letter of the agreement. In no better way can the magnitude of the undertaking be made manifest, than by giving a few statistics in regard to the tunnel. It is 25.031 feet in length, and nearly midway in its length rises the central shaft, which terminates near the top of Hoosac Mountain, and is 1,040 feet deep. This shaft is oval in form, the major axis of 27 feet being coincident with the line of the tunnel, and the minor axis being 15 feet. It has been said that the sections on either hand of the central shaft were of nearly equal length. The section opening at the eastern portal is 12,837 feet long, while that to the west of the central shaft is in length 12,194 feet.

The First Plan for a Tunnel.

It was as long ago as 1825, that a Board of Commissioners was duly appointed to consider the practicability of building a canal from Boston to the Hudson river, having its western terminus at the place where the great Erie Canal, the pride of the time, emptied its waters into the river. After a careful examination into the various routes which were proposed, with more or less eagerness and zeal, by the rising politicians of the Commonwealth in that day, the Commissioners submitted a report in favor of following the course of the Deerfield and Hoosac rivers, and of passing through the Berkshire hills by means of a tunnel through the Hoosac Mountain.

But just as the subject began to be discussed, the railroad began to be heard of; so that, in the interest and enthusiasm which was felt for the new mode of conveyance, all the labor and all the arguments of the Canal Commission were forgotten.

The First Charter.

With the history of the railroad in this Commonwealth our description has to do only so far as it is connected and interwoven with the history of the tunnel; but it may be said that the tunnel line had its birth when the Vermont and Massachusetts Railroad Company was formed. One year before, the Fitchburg road had been built in a thorough and substantial manner in the short space of less than two years. Only six years after the completion of the last-named road, a charter was granted by the General Court, acting in obedience to popular demand, to the Troy and Greenfield Railroad Company, giving the corporation the right to build their line from near Greenfield to the State line at Williamstown, and there to connect with a railroad which should be built from Troy to that point. There was no mention made in the charter of a tunnel through Hoosac Mountain, but it is quite evident that the tunnel formed a part of the plan.

Application for State Aid.

Subscriptions to the capital stock were few and far between, and, before a year had elapsed, the corporation determined to apply to the General Court for a loan of the State's credit; and it was in the Legislature of 1851 that the contest on the tunnel began. There were to be found many who remonstrated against the State's taking any action in the matter; but, after mature deliberation, the committee came to the conclusion that $1,948,557 was a sum more than enough to finish the tunnel, and that allowing plenty of time for accidents and hindrances, 1556 working days would amply suffice for the accomplishment of the work if no shaft were sunk. With a shaft, the committee were sanguine enough to suppose that the tunnel could be completed in 1954 days. The discussion was not very brisk, however, until the plan for State aid had been broached; and the editor of every country paper in the Commonwealth had then something to say on the subject. In the Legislature the contest was slowly but fiercely carried on; and on the 12th of May the project for a State loan was defeated by the strong vote of 108 in favor to 237 opposed. In 1852 the vexed subject was dropped, the managers of the Troy line doubtless feeling that there was nothing to be expected from the General Court of that year; but, in the year following, the corporation again appeared at the bar of the Legislature, asking for a loan from the State. On the 26th of April, in the closing days of the session, the bill was passed to be engrossed by a vote of 143 to 96. But so much opposition was manifested, not only by some of the most influential journals of the time, but by men whose opinion was powerful in influencing the popular mind, that, when the measure came up for final action, the loan bill was lost and the tunnel postponed for still another year. In the following year, however, the advocates of State aid renewed the contest, and had the satisfaction of finding their labors at last crowned with success.

The First Loan Granted.

The law providing for the first Hoosac Tunnel loan was drawn up with great care, and even the opponents of the tunnel conceded that no pains were spared to keep the credit of the Commonwealth untarnished in the transaction, which it was confidently predicted would make the State bankrupt before 20 years. The sum of $2,000,000 in sterling bonds was fixed upon as the amount to be loaned by the State, and to be paid in instalments of $100,000 each. In order to properly provide for the State's credit, various regulations were made. Under these restrictions the enterprise did not attract capital, and subscriptions were sent in with alarming infrequency.

The Earliest Contracts.

But nevertheless the contract for the road, which had hitherto existed only on paper and in the imaginations of its projectors, was awarded in spite of the scarcity of funds; and the work of constructing the line, tunnel and all, was given in 1855 to Messrs. E. W. Sewell & Co. of Philadelphia, for the sum of $3,500,000, which was subsequently annulled. But hardly had two months elapsed when negotiations were entered into with the firm of Hermann Haupt & Co.; and in July, 1856, the first contract was made with those gentlemen, who seem to have had great confidence in the success of the enterprise. By the agreement, Messrs. Haupt & Co. were to build the road and tunnel for the sum of $3,880,000, the money to be obtained as follows: From the Commonwealth, $2,000,000; in mortgage bonds of the company, $900,000; in capital stock, $598,000; in cash, $382,000. Six thousand shares of the stock were to be taken by Haupt & Co., no cash assessments to be laid thereon, but the amount due to be paid in work, as stock credits. After another application for State aid, in 1857, the Legislature voted to send a special committee to the tunnel itself, and to make a report on the condition of matters. This was the first of the many official visits which legislators have, from time to time, paid to the romantic neighborhood of Hoosac Mountain. In a month's time, the committee made a report, showing that 621 feet had been pierced at the eastern portal, and 185 feet at the west of the mountain; and thereupon arose the most famous controversy in the history of the Hoosac Tunnel. On the question of modifying the requirements of the loan act, there were presented, finally, two reports. The majority and the minority — the last consisting of one member — in the committee agreed that the rigor of the act should be relaxed; and all but one obstinate gentleman were willing to be liberal with the company, and give them every reasonable opportunity to complete the work which they had in hand. On the 14th of May, therefore, a bill was passed by both houses of the Legislature by heavy majorities, and the matter was vetoed by Governor Henry J. Gardner of "Know Nothing" fame.

A Legislative Investigation.

In 1860, after various vicissitudes, it occurred to the Legislature that it would be well for a committee to pay a visit to Hoosac and to see how the work was going on. That there was any suspicion that the corporation and the contractors were not doing all that was required by law there seems to be no evidence. But the committee found a state of things around Hoosac Mountain which they little anticipated. The State Engineer, in his report on the condition of the road in 1858, said that "the total length of rails laid was seven miles, 138 feet," but when the committee came to enquire into the subject, it was found that the road was not completed for that distance, but that there were several places where the road was not passable. In January, 1860, the third instalment of scrip had been paid, it being certified that the road had been graded three miles further. But the committee found out that in this length of three miles there were fourteen gaps, every point where a bridge, culvert or cutting was to be made, had been left untouched. This statement was submitted by the committee to the Legislature, without note or comment, and the matter was remanded to the special committee on the tunnel.

The Haupt Contract Annulled.

The company struggled on with the work during the year, and for nearly half the year following; until in July, 1861, the State Engineer, finding that the contractors failed to do the prescribed amount of work, refused to certify their bills, and the Troy and Greenfield corporation, after so many years of hard work, laid down the Hoosac Tunnel enterprise, and so the history of the project under private management came to an end. For two years nothing was done, though Governor

Andrew, in his message to the Legislature, in 1862, recommended the General Court to take prompt action. The whole subject was, early in the session, referred to a special committee, who gave a patient hearing to all parties interested, and, in March, presented a report, with whose conclusions each member of the committee agreed. This report regarded the faith of the State as pledged to an investment of at least $2,000,000, on account of the tunnel, and saw no reason why the State could not undertake the work single-handed. The action of the State Engineer, sustained by the opinion of the Attorney-General, the Hon. Dwight Foster, was declared to have resulted from a misapprehension of the meaning of the legislative act of 1860. The committee, on the assumption that "no one contemplated aid from the State to an extent exceeding $2,000,000," recommended that the road east of the tunnel should be completed by the State, at an expense not greater than $195,000; and that the work of tunnelling should be paid after a "red-tape" system, which need not here be given, as the bill never passed the Senate, and was never heard of in the House. The measure was lost on the 21st of April, by the casting vote of ex-Governor Clifford, and in its place a measure was offered by the Hon. William D. Swan, of Dorchester, which provided that the State should at once possess itself of the road and tunnel, and proceed to finish it, in any way thought best; and, when done, to run it or leave it, as might be thought advisable. The bill passed both houses, with an amendment limiting the entire expenditure to be incurred by the State, under this and all previous bills, to $2,000,000, and Governor Andrew approved the measure on the 28th of April.

State Commissioners Appointed.

The State having, in spite of a vigorous opinion, committed itself to the completion of the tunnel, if the project was a thing to be accomplished by mortals, three Commissioners,

Messrs. John W. Brooks, S. M. Felton and Alexander Holmes, were appointed, who entered upon their work early in May. The directors of the Troy and Greenfield road, voted in August, to surrender the property of the corporation to the State, and on the 13th of October, after some necessary delays, the President of the road, and the firm of Hermann Haupt & Co., submitted to the act of the State and relinquished all right, title and interest in the tunnel. Nothing had been done on the tunnel since the Haupt contract had been cancelled in 1861; and the mountain, when the Commission resumed operations in August, 1863, had been pierced only at the east end to any extent, and there only to a distance which was less than one-tenth of the smallest estimate of the length of the shaft when completed. The five years of tunnelling, under the State Commission, forms an interesting chapter in the history of the great Hoosac bore, and makes a record of steady and persevering work, in the face of every obstacle, of which those concerned may well feel an honest pride, although they can boast of but little actually accomplished.

Compressed Air as a Motive Power.

Up to this time all the actual work had been accomplished by hand power. The much vaunted "boring machines," which were to run straight through the mountain, had left only their unpleasant memories behind them. But just at this time the use of compressed air as a motive power began to be discussed. In the previous year's session, opinions had been given in favor of the project; and the Commission now thought in what way such motor power might be attained. After much discussion it was decided to dam the Deerfield River at a point about three-quarters of a mile from the eastern portal, and bringing the water from thence to the tunnel by a canal, thus securing a fall of about 30 feet. The project provoked a storm of opposition, and was characterized as a great and most costly blunder. But the Commissioners stoutly maintained its utility, and asserted that it would not only furnish a constant supply of compressed air at the east end, but that its benefits could reach the central and new west shaft workings, with only the expense of carrying the pipes over the mountain. The result has at any rate disproved the charge of total inefficiency which was brought against the Deerfield dam project by its many opponents, for sufficient power was furnished almost constantly for the working at the east end, though the plan never worked at the the opposite side of the mountain.

The Results Accomplished in 1864.

Work went on steadily through the Winter of 1863-64; and in September of the latter year the following report was writ by State Engineer Doane, showing the progress of the undertaking: At the west end, a new portal had been cleared, and the machinery was in readiness to begin the work of tunneling. At the shaft on the west, where nearly 60 feet had been pierced by Haupt & Co., the west working had gone forward 45 feet, and the east working 128 feet. The east end shaft had not made any progress, but the work of enlargement had been carried on with some success. Work on the central shaft had only been fairly begun; in the following January, 1865, it had been sunk only to the depth of 74 feet. Such was the record up to the Spring of 1865. In March of that year, the work of cutting down the breast of the work in the tunnelling from the east was brought to a conclusion, and henceforward, progress was made from the bottom, instead of the top of the tunnel. The work had made a statisfactory progress, the gain in one year being 239½ feet. At the central shaft progress had been made to the depth of 127 feet, the shaft being lighted by naphtha gas in October.

The Years 1866 and 1867.

The year 1866 chronicles some important events at the tunnel. In June, the new machine drills for the use of compressed air at the eastern end were introduced, the machinery of the water works having been in readiness the previous January. At first the progress made was less than what had been accomplished by hand-power; but certain improvements were soon made, which resulted in the success of the experiment, so far as the east end was concerned. At the west shaft, nitro-glycerine was introduced, and brought into use everywhere, except at the east end. The total amount pierced was: East end, 3473; west shaft, eastward, 1042; central shaft, 377. The total length of headings, excepting at the west end, was 4813 feet, of which 1325½ feet is to be credited to the work of the 13 months ending December 1, 1866. During 1867 further improvements were made. The water had increased so much at the west shaft, east heading, that work was suspended until June, when a new contract was made with Mr. Farren, for the work on the west end of the tunnel, requiring brick arching; and Messrs. Dall, Gowan & Co. contracted to undertake the work at the east end and central shaft, during the months of August, September and October. Under these gentlemen the work was carried on throughout the year, with the following result: The west heading, from the west shaft was carried forward 313 feet, while the gain at the eastward heading was 252 feet; at the east end, an advance had been made in 13 months of 1235 feet, and the central shaft had been sunk to a depth of 206 feet further than a year before. Under Mr. Farren's contract, the brick tunnel from the west end had been carried forward to a length of 414½ feet, with an additional heading of 297 feet. A summary of the year's work, therefore indicates a total length of tunnel and heading of 7324½ feet, of which amount 2511½ feet were gained by the labor of the past 13 months.

1868 — A Legislative Battle.

But the battle which was waged on the subject in the General Court of 1868 was long and bitter. The enemies of the tunnel project spared no effort to induce the Legislature to abandon the undertaking on which had been spent so many millions of dollars. On the 15th May, the Special Committee on the Tunnel reported a bill appropriating $250,000 for the completion of the railroad, $600,000 for work on the tunnel during the year, together with $350,000 for interest on debt; and, moreover, authorizing the Governor, by and with the advice of the Council, to enter into a contract for the completion of the road and tunnel, and to dispose of the State's interest in the enterprise on such terms as appeared advisable. On the 11th of June, after one bill had been passed by a small majority and found to be illegal, a second supplementary bill prevailed in both houses and received the Governor's signature. The bill embodied the features of the first section of the bill submitted by the Tunnel Committee, but was amended by Mr. Packard so as to authorize the making of a contract to complete the entire work, provided it could be done in seven years, at an expenditure of not more than $5,000,000, and to preclude any expenditure by the State in work not done under contract after October 1. Other amendments made up the bill as it passed finally to be a law, requiring "satisfactory guarantees" from the contractors, and withholding at least $1,000,000 from the contract price until the enterprise was completed. The record of work accomplished when operations were entirely discontinued, in October, shows that in nine months the east end heading had advanced 574 feet; amount pierced at the west end, west shaft and well No. 4 workings, 1504 feet, and in all 2088. There remained to be completed, 15,693 feet of tunnel and 457 feet of depth in the central shaft.

The Shanley Contract.

When the proposals for the contract were sent into the Governor and Council, it was found that the estimates varied from $4,027,780 to $5,378,354. Of the twelve bids which were made, only four were within the limit of $5,000,000, fixed by the Legislature to defray the expense of tunnelling, and to liquidate outstanding liabilities which amounted to about $250,000. Messrs. Francis and Walter Shanley, of Canada, made a tender to undertake the enterprise at $4,623,069. Although this was next to the highest bid made by the four whose proposals fell within the necessary limit, the Legislature had not required that the contract should be given to the lowest bidder, and in consideration of the deposit of public securities to the amount of $500,000 as a satisfactory guarantee, as well as in view of the long experience and excellent reputation of the firm, Governor Bullock gave the contract to the Messrs. Shanly, after long consultation with the Council. The contract was signed on the 19th of December, 1868, but work was not actually begun until the 29th of March, in the year following, when operations were resumed at the heading at the east end. The record of work done up to the close of the year showed that the total length of the east end heading was 6522 feet, and of the west end, 4505; in all, 11,027 feet out of 25,031. In 1870, steady progress was made, the central shaft completed to a depth of 1028 feet, two feet less than required by contract, and work commenced on a heading in each direction. In August, sixteen feet had been built in either direction, exclusive, of course, of the width of the shaft. The record for the year is to the effect that 1514 feet had been pierced at the east end, 1203 feet at the west end, and 60 and 80 feet east and west of the central shaft; thus, during the year, 2864 feet of tunnelling had been pierced, leaving only 11,140 feet to be accomplished. In 1871, a large flow of water prevented progress to any great extent in the central shaft, and it was generally thought that the contractors would be unable to complete the tunnel in the given time. The aggregate of progress showed a gain of 1743 feet at the east end, 1380 feet at the west end, and 430 feet east and west of the central shaft; in all, 17,446 feet accomplished, and divided as follows: east end, 9779 feet; central shaft, east, 337 feet; central shaft, west, 240 feet; west end, 7090. The contractors worked so faithfully during the year 1872 that the average required by contract was much exceeded, excellent progress being made in every direction except at the central shaft, west heading, where an advance was made of only 119 feet. On December 12th, a junction was made between the east end and central shaft workings. It will be, perhaps, remembered, that when the lines met, a varation of but five-sixteenths of an inch could be discerned.

The Mountain Pierced.

In 1873, work was pushed so vigorously that there was remaining to be done, November 1, 1873, 242 feet. The record thereafter stands:

November 8, advanced from central shaft, west,	31
November 8, advanced from west end east,	25
	56
November 15, advanced from central shaft west,	42
November 15, advanced from west end east,	33
	75
November 22, advanced from central shaft west,	41
November 22, advanced from west end east,	29
	70
Total,	201
Remaining to be pierced, November 23,	41

Leaving, on November 23, forty-one feet to be completed, which work was completed on that day in the presence of a large number of State officials, railroad men and journalists, who passed through the opening in the wall of rock, and made the tour of the Hoosac Mountain, underground. From that time only the work of blasting out the tunnel to its full size, laying the iron and completing the road at either end, remained; and though this was in itself a considerable task, and involved a much greater outlay of time and money than any of the ardent friends of the enterprise foresaw, it has all been successfully accomplished, and the passage of the tunnel forms the most notable feature of one of the principal routes from Boston to Saratoga and the West.

North Adams and its Surroundings.

Emerging from the gloom of the tunnel, a short run over a descending grade and around a rather sharp curve, brings us to the pretty village of North Adams, situated at the forks of the Hoosac River, in a perfect amphitheatre of hills, from which views both grand and charming are gained. Aside from its importance as a railway junction,—the Hoosac Tunnel line and a branch of the Boston and Albany here meeting— North Adams has no inconsiderable prosperity as a manufacturing village. Here are 20 cotton and woolen mills, power for which is furnished by the two branches of the Hoosac, which unite near the centre of the village, and several large shoe-factories. In one of these latter the first experiment in the Eastern States with "Chinese cheap labor" was tried by Mr. Sampson in 1870, some 75 Mongolians being imported from San Francisco. The population of North Adams is about 13,000 of which some 5000 souls are employed in the various factories. Here are several neat churches, a fine high school-house, many elegant residences and two hotels, the Ballou House and the Berkshire Hotel. There are many points of natural interest in and about North Adams. Excursions are plenty and easily taken, for the ascent of Mount Greylock, for visits to the natural Bridge — one mile east of the village,—where Hudson's brook has worn a passage 30 rods long and 15 feet wide through the solid marble, which stands an arch 30 to 60 feet high above — and where Hawthorne was fond of straying and musing during a summer spent at North Adams in 1838; to the cascade on Notch Brook- one and a half miles from the village, where the water leaps down 30 feet; and best of all, the drive over the Hoosac Mountain. The ride is eight miles in length. The west peak is scaled by a succession of zigzags constantly rising, and from its summit is gained a splendid view of the neighboring villages, the Hoosac Valley, Greylock and the Vermont hills. Then a swift descent brings us to the "saddleback" or plateau between the summits, where we pass the central shaft of the tunnel. Then the eastern summit is climbed, and the view presented is grand and majestic, justifying fully Hawthorne's eloquent description:

A noble view is obtained from this point, above the romantic gorge of the Deerfield river to Wachusett Mountain, and beyond it the blue and indistinctive scene extended to the east and north for at least sixty miles. Beyond the hills it looked almost as if the blue ocean might be seen. Monadnock was visible like a sapphire cloud against the sky. The scenery on the east side of the Green Mountains is incomparably more striking than on the west, where the long swells and ridges have a flatness of effect. But on the eastern part, peaks one to two thousand feet high rush up on either bank of the river in ranges, thrusting out their shoulders side by side. Sometimes the precipice rises with abruptness from the immediate side of the river; sometime, there is a valley on either side; cultivated long, and with all the smoothness and antique rurality of a farm near cities, this gentle picture is strongly set off by the wild mountain frame around it. I have never driven through such romantic scenery, where there was such variety and boldness of mountain shapes as this; and though it was a sunny day, the mountains diversified the view with sunshine and shadow, and glory and gloom.

Proceeding westerly by Troy and Boston Railroad from North Adams, a ride of about one mile brings us to the crossing of the highway to Williamstown and the Hoosac river. Here a small elm can be seen in a meadow, only a few rods from the track, marking the site of old Fort Massachusetts, one of the cordon of works built by the colonists in 1744 to guard the frontier. From this point, railroad, river and highway run amicably side by side through a narrow defile to Williamstown, a lovely hamlet nestled in the lap of the mountains, and noted as the site of Williams College, as well as for its beautiful mountain and valley scenery. South of the village is Mount Hopkins, 2,800 feet high, which is often ascended for the sake of the view from its summit. The Hopper, also south from the colleges, is a vast gulf bounded by Greylock on the east, Prospect Mountain on the north and

Bald Mountain on the south, and abounding in fine cascades and noble scenery. Snow Glen, where snow always remains, and Flora's Glen, where in 1812, William Cullen Bryant, then 18 years old, and a student at Williams College, wrote his "Thanatopsis," are places of interest in the vicinity. Two miles north of Williamstown is the famous Sand Spring, beneficial in cutaneous diseases, where is located Greylock Hall, a large new hotel.

Into New York State.

From Williamstown, which is the extreme northwestern town of Massachusetts, our course is northwest, skirting the rocky town of Pownal, the southwest corner of Vermont, to Petersburg, the first town in New York State. Thence passing Hoosick Junction, where we cross the Harlem Extension railroad, we are soon whirled into the Union depot at Troy, N. Y. Troy is the capitol of Rensselaer county, is the head of navigation on the Hudson, has about 50,000 inhabitants, is noted for its iron foundries, stove manufactories, bell foundries, prosperity, Willard Female Seminary, fine residences, paper collar works, breweries, laundries, and several other things, too numerous to mention. There are several handsome streets, bordered by elegant residences, and many fine churches, but the chief glory of Troy is its thrift, springing from its many and varied industries, its commerce on the Hudson, its wide-spread tributary country, and its excellent educational advantages. On Mount Ida, a tall hill overlooking the city on the east, is St. Joseph's Theological Seminary, a Roman Catholic institution for the education of priests; the Rensselaer Polytechnic Institute, founded in 1824 by Patroon Van Rensselaer, has a world-wide fame, while the Female Seminary, founded in 1821 by Mrs. Emma Willard, has been pre-eminent for many years, and numbers over 7000 alumnæ. West Troy, just across the Hudson river, which is spanned by a steam ferry, is a busy suburb, with its manufactories, over 40 in number, and its Watervliet Arsenal, one of the largest of Uncle Sam's establishments. From Troy, morning and evening lines of boats run to New York city, stopping at Albany and other river ports; trains leave the Union depot for Albany, New York and the West by New York Central and Hudson River Railroad; for Albany, Saratoga and Rutland by Rensselaer and Saratoga Railroad, and for Boston and New England generally by the route over which we have come. Leaving the Union depot by the Rensselaer and Saratoga train, a ride of six miles brings us to Albany Junction, where a branch from Albany, 12 miles distant, joins us, and we continue our course to Saratoga by rail as already described.

A Rail Car Flirtation.

The pleasures of a rail car ride to Saratoga are sometimes heightened, and its minor discomforts alleviated by an innocent flirtation, if opportunity serves, as graphically described in verse by Mr. J. Cheever Goodwin, of Boston:

> Laughing eyes and pouting lips,
>   Dainty waist and taper,
> Sat she just across from me,
>   Reading morning paper.
>
> Golden hair, befrizzled, in
>   The style that's most in vogue, her
> Ticket in her hand explained
>   She went to Saratoga.
>
> Much I pondered as I sat,
>   Scanning her with caution,
> Whether the inviting chance
>   I should welcome or shun.
>
> She was all alone, you see;
>   Should I greet her boldly?
> Would she welcome my approach
>   Or receive me coldly?

"Nothing venture, nothing have,"
    Think I, as gets lost her
Ticket on the floor,—and straight
    Find it and accost her.

"Travelling alone?." I asked,
    "Yes," she answered sweetly;
"But a friend will meet me, though,"
    Added she discreetly.

Thence progressed I rapidly,
    Shared her tempting luncheon.
She, though circumspect, did not
    Any gentle fun shun.

Much I wished, as on we sped,
    Toward our destination,
That the train would practice some
    Slight procrastination.

All in vain, too soon, alas!
    Reached we Saratoga;
There was waiting for her,—well,
    Set him down an ogre.

All this happened yesterday;
    Would you know the sequel?
Listen to a moan, so sad,
    Never was its equal.

I, to-day, at each hotel,
    Diligently sought her,
Hoping she might prove to be
    Some one's only daughter.

And I found her, well-a-day!
    Only child she *may* be;
Certain 'tis she's married, and
    Rejoices in a baby.

CHAPTER IV.

Saratoga and Its Attractions.

THE village so famous in the annals of fashion is situated in a valley running nearly northeast and southwest, beneath the surface of which is nature's most potent laboratory, whose preparations come to the surface in scores of places within a few miles. The Ballston Spa, so famous a half century ago, are in the same valley at its southern end. Yet though the extent of territory in which these healing mineral springs have been found to exist is quite limited, yet within its extent and out of its scores of springs, no two alike are in chemical analysis or in therapeutic effects. And this is the greatest wonder of all, that such a wide diversity of composition should exist in springs close together, — in some cases within the same enclosure. Of all the medicinal waters of Europe, the Spa of Belgium, the Seidlitz of Bohemia, the Selzer, Baden Baden and Aix of Germany, and the Chelten-

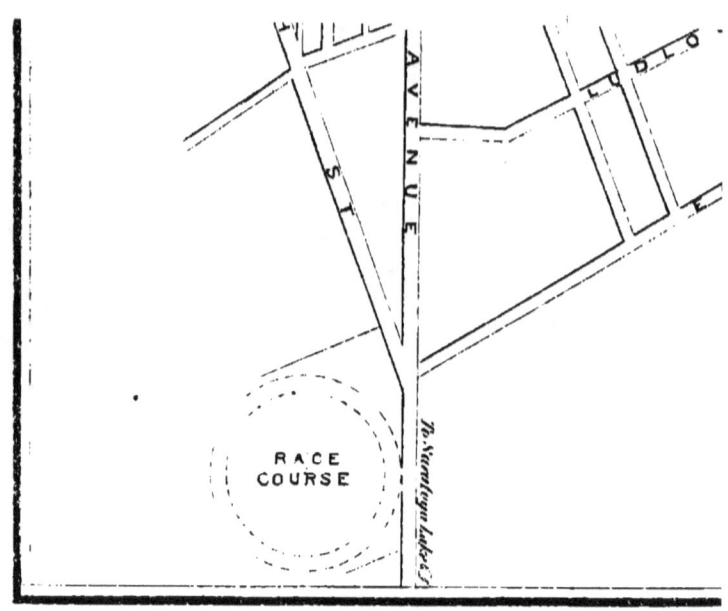

W.H.FORBES,CO

LLAGE

SPRINGS

PARK

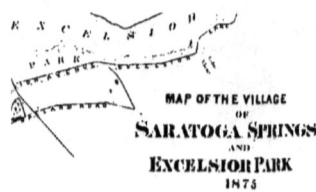

ham, Bath and Harrowgate of England, all are inferior in variety and power to the inexhaustible and health-giving springs of Saratoga. The site of the village is in the hear of the valley, itself some 300 feet above the sea, at the foot of the Kayaderosseras Mountains, which form the water-shed between Lake Champlain and the St. Lawrence. The healthfulness of the location and its convenience to railroad facilities have made Saratoga a favorite Summer residence with wealthy people, who can afford to keep their own cottages or villas here, and who are yearly beautifying the place by the fine buildings they are erecting, and the handsome grounds they are laying out. The principal street is Broadway, which nearly follows the course of the valley, and on which the principal hotels front. Broadway is a fine, straight, wide avenue, and when crowded of an afternoon, with the elegant equipages of Saratoga's Summer residents, slowly moving up and down, as if on dress parade (which the ladies *are*), the sight is beautiful and full of life. Many of the most noted springs are on or near Broadway, and here the chief business of the place is transacted. The first knowledge of the healing qualities of the waters hereabouts was gained by the Indians, who in 1757 brought to the valley on a litter Sir William Johnson, Bart., a friend and patron of the amiable red men, and in a few weeks he was restored to health. The spring at which this cure was effected was the High Rock, which is still famous, and is distinguished from all other fountains by the mound of calcareous tufa, seven feet in diameter in the base and half that distance in height, which the waters have deposited in the course of centuries. This mound is in the form of a low cone, and from a hole some six or eight inches in diameter, at its centre, the clear and pleasant water bubbles np. Naturally the superstitious savages harbored the most profound reverence for this healing spring, and the early settlers shared in their appreciation. Probably the next spring discovered was the Congress, which was first observed by a party of hunters, one of whom was ex-Congressman Gilman, (in honor of whom it was named Congress Spring), in 1792, as they were strolling along the banks of the little creek, into which its waters trickled. Subsequently, being tubed, its flow was greatly increased, and it now is the most noted and one of the most prolific (if the word be applicable), of the healing waters. From time to time, within the present century, other springs have been discovered or developed by boring, until now, the valley for several miles is honeycombed with the tubes of mineral springs, bearing all sorts of names and possessing all sorts of properties. To enumerate them all in proper order, we must suppose ourselves setting out to drive through the village, and visit each in turn, and as the most natural course will begin at the very centre of the village, where, within a few rods of the three largest hotels, is the beautiful and world-famous

Congress Park.

This public pleasure ground, which is always the first place of interest visited by the newly-arrived, is a pretty enclosure of a few acres, of nearly triangular form, bordered by Broadway, Union Avenue and South Circular street, naturally diversified in contour, and covered with a handsome growth of trees. The proprietors, "Congress and Empire Spring Company," have laid out graveled walks, undulating lawns and secluded copses, and have placed statuary at various points, besides stocking the park with deer, which are very tame and roam freely, and building handsome structures over the two springs, Congress and Columbian, which stand near the corner of Broadway and Union Avenue. The Doric colonade in the left of the illustration covers the Congress, and the Grecian dome to the right surmounts the Columbian. The park is the resort of thousands of people, residents and visitors, daily, and its shaded walks are a favorite stroll. Could the arching trees but speak, they could tell many a romantic tale of sweet flirtations and delightful love-making which they have witnessed. Presumably, most people come

CONGRESS SPRING AND PARK, AND COLUMBIAN SPRING, 1873.

hither to drink the waters —some take half a dozen glasses at a dose,— but the winding paths, the sylvan nooks and the companionship of the *dears* are so productive of tender emotions, that many who come to drink remain to "spoon."

The Congress Spring.

The properties of the Congress water, which as well as the most celebrated, is among the most efficient of the Saratoga waters, and which, in bottles, can be found almost the world over, are pleasantly, but effectively cathartic. The taste is agreeable to most people, and the cool, sparkling draught is certainly one of the "easiest to take" of all medicines. The carbonic acid which causes its effervescence also imparts a vivacity to the water, which resembles that of soda water. In small quantities after meals, the Congress water gives tone to the stomach and clearness to the head, especially agreeable to those whose business tends to mental exhaustion and physical torpidity. The water is useful as a preventive of feverish and biliary disorders, headaches and the like, and is freely drank by the residents. To secure the fullest effect the water should be drank before breakfast, in quantities easily regulated by experience, its effect being aperient without subsequent reaction or languor, and a gradual diminution of the quantity leaves the bowels regular, and the general health and spirits greatly improved. As an alterative or tonic, the water should be taken sparingly through the day with brief intervals, and the effect is suprisingly beneficial. In cases of chronic dyspepsia, diarrhea, jaundice and diseases of the kidneys, the water properly used will remove the evil accumulations from the bowels and stimulate the organs to the normal discharge of their functions. As a remedy for the disorders of sedentary life, constipation, with all its resultant ills, and the disorders occasioned by "high living," the water has a most salutary effect. The operation of the Congress water, though certain and thorough, is free from griping pains, and its after effects pleasantly different from those of drastic purgatives. In submitting a new analysis of this water, which is given elsewhere, Professor C. F. Chandler, Ph. D., of Columbia College, says: "A comparison of this with the analysis made by Dr. John H. Steel in 1832, proves that Congress water still retains its original strength, and all the virtues which established its well-merited reputation." It should be remembered that the water of this spring is sold in *bottles only*. What purports to be Congress water, for sale on draught in various places throughout the country, is not genuine. The artificial preparations thus imposed upon the public may have a certain resemblance in taste and appearance, but are frequently worse than worthless for medicinal purposes.

The Columbian Spring.

This spring, which is within a biscuit-toss of the Congress and owned by the same company, was opened in 1806 by Gideon Putnam. The water issues from the natural rock, about seven feet below the surface of the ground, and is protected by heavy wooden tubing. It is the most popular spring among the residents of Saratoga. The escaping bubbles of free carbonic acid gas give to the fountain a boiling motion. Large quantities of the gas can easily be collected at the mouth of the spring at any time. It is a fine chalybeate or iron water, possessing strong tonic properties. It also has a diuretic action, and is extensively used for that purpose. The water is recommended to be drank in small quantites during the day, generally *preceded* by the use of the cathartic waters taken before breakfast. It is put up in half-pint bottles by the company, and is especially valuable in liver complaints, dyspepsia, erysipelas, cutaneous diseases, chlorosis and many female complaints.

The Empire Spring.

The same company owns the Empire Spring, which is situated on Spring avenue, at the head of Circular street, and

EMPIRE SPRING AND BOTTLING HOUSE, SARATOGA, N. Y.

near the base of a high limestone bluff, in the northerly part of the village, a few rods above the Star Spring, and about three-fourths of a mile from the Congress. Mineral water was known to trickle down the bank at this point ever since the land was cleared of its primitive shrubs. It was not till the year 1846 that the fountain was tubed. Tne proprietors have surrounded it with shade trees, built a pavilion over it and erected a large bottling house close at hand. The water much resembles that of the Congress Spring, but is more active, owing to a greater amount of magnesia. The Empire is highly esteemed for the treatment of obscure and chronic diseases, requiring alterative and diuretic remedies. It is also valuable as a preventive of intermittent, billous and gastric fevers, dysentery and liver complaints, rheumatism, gout and cutaneous disorders, etc. The same rules apply to its use as have been given for the Congress water. Some systems take more kindly to the one, and some to the other water, but the general effect is much the same. Another celebrated water is that of

The Star Spring,

which is located near the Empire, on Spring avenue; Star Spring Company, proprietors, J. W. Dane, President. Under the name of President Spring, and afterward Iodine Spring, the fountain now called the Star has been known for nearly a century. It was first tubed in 1835. In 1865 the Star Spring Company was formed, and in the following year the spring was retubed under their direction. In 1870 they erected the finest bottling-house in Saratoga. Great care is taken to preserve the spring in a pure condition and perfect repair. The water has become immensely popular in New England, and throughout the United States and Canada. The proprietors of this spring, feeling the need of some method of transporting the water in bulk, to avoid the heavy cost of bottling, and the heavy freight upon the same, commenced in 1866 to send the water in barrels made of rock maple. This method proved a failure, as it was impossible to confine the gases in wood, and impossible to prepare the wood in a manner not to impart to the water its peculiar taste. They then prepared the tin-lined barrels (patent dated November, 1867) which proved a success. These barrels are used to convey the water to all parts of the country. It is then drawn into fountains, and charged lightly with gas to restore it to its original condition, and is dispensed by the glass, and is as palatable and effective as at the natural fountain. This method has become very popular where known. For commercial use, the water is sold in cases of quarts and pints; and besides, owing to the large amount of gas which is finely incorporated with the water, the company are enabled to supply families with it in kegs of 15 gallons, in which the water keeps as well as in bottles, and at one-fourth to one-sixth the cost. This method seems to give entire satisfaction, and is fast coming into general use. The price to druggists, in bulk, is 20 cents per gallon; to families, four dollars per half barrel; to the trade, in cases, at 21 dollars per gross for pints, and 30 dollars per gross for quarts. The large and pleasant office in the bottling-house is adorned with flowers, shrubs and rare exotics of great beauty.

Visitors will find here ample accommodations for rest and recreation, as the office is open to all.

The Star water is mildly cathartic; has a pleasant, slightly acid taste, gentle and healthy in its action, and yet powerful in its effects. It is far more desirable for general use as a cathartic, than the preparations of the apothecary. The Rev. Dr. Cuyler, in one of his peculiarly charming letters, gives the Star water preference over all others as an active and efficient cathartic. While the immediate effects of the Star Spring are cathartic, its remote effects are alterative; and these, after all, should be considered the most important, as the water thus reaches and changes the morbid condition of the whole system. In this part of the village, and near the Star, is the famous High Rock Spring above mentioned.

In the same vicinity and under the same proprietorship, is the Saratoga "A" Spring. In 1865, Messrs. Western & Co. purchased the property, and sunk a shaft 12 feet square, to the depth of 16 feet. The surface above the rock consists of bluish marl, similar to that found all along this mineral valley. A tube, in the usual form, was placed over the spring and clay was used as packing around it. In the spring of the next year, the fountain was more perfectly secured by a new tubing; and the water was bottled, and shipped all over the country. Near by is the Red Spring, long known and valuable in cutaneous diseases, scrofula, dyspepsia, and the Seltzer Springs, which produces a pleasant beverage, much like the imported seltzer water, and used extensively for mixing with still wines, etc. Returning towards the centre of the village, we come to

The Pavilion Spring.

This has for more than thirty years been favorably known. It is central in position, and, with the neat park around it, is a pleasant place of resort. Church street bounds the park on the north, Spring avenue extends northward, and the elegant structure which surmounts the fountain is but a few steps from the street on either side. The United States Spring is under the same colonnade, and the water is pleasantly cathartic. Passing the Pavilion and turning down Putnam street, we come to the spring of the same name, which is chiefly used for bathing purposes, the water being a strong chalybeate, or iron tonic. The Hathorn Spring comes next, on Spring street, in rear of the Congress Hall. It was tubed in 1869, and is the most active cathartic to be found at Saratoga. The Hon. H. H. Hathorn is proprietor, and the water is bottled, as well as extensively drank on the spot. The Hamilton Spring, on the opposite side of Spring street, is principally diuretic in action. Passing Congress Park, we come on the other side of Broadway, to the beautifully shaded grounds of

the Clarendon Hotel, in which, under a pavilion, is the Washington Spring, often called Champagne Spring, from its peculiar effervescence. It is tonic and diuretic in its action, and strongly impregnated with iron. It was tubed in 1806. The Leland Spring is in the same grounds. Having thus briefly noted the Springs near the centre of the village, we will next take a trip to some of the outlying fountains. Of these, the most celebrated is the Excelsior Spring, about a mile east of Broadway, and a little north of the Empire Spring, in a most romantic and beautiful dell, formerly known as "The Valley of the Ten Springs," but which has been christened, "Excelsior Park." In the same vicinity are the Union (formerly the Jackson), a mild cathartic, the Minnehaha, the White Sulphur (used for bathing), and the Eureka Springs.

The Geyser Spring.

This, with the neighboring Triton and Champion springs, are the spouting springs of Saratoga, and are about a mile and a half south of the village on the Ballston road. All are artificial, having been successively bored in 1870, 1872 and 1871 respectively. The Geyser Spring is in a building which for some years was used as a bolt factory, and the proprietors of which sunk a shaft in hopes of finding water. The boring is 140 feet deep, $5\frac{1}{2}$ inches in diameter; sixty feet of the distance bored was through limestone. A tube was fitted to the boring and connected with a pipe so that a constant stream is playing to the height of fifteen or twenty feet. This perpetual fountain is much visited as a curiosity. The water is singularly cold, being only 14° Fahrenheit above the freezing point. In mineral ingredients this spring is the richest of the Saratoga waters that have yet been analyzed. As a medicinal agency its effects are marvelous. Testimonials from all quarters are received bearing witness to its wonderful cures of diseases, especially in cutaneous diseases or any of the various phases of scrofula. It is used with telling effect in kidney disease, liver complaint, dyspepsia, biliousness, rheumatism, acidity of stomach, constipation and piles. It is a delightful beverage, and when taken as a

Analysis of the Waters.

	High Rock Spring.	Pavilion Spring.	United States Spring.	Hathorn Spring.	Congress Spring.	Geyser spouting well.	Union Spring.	Empire Spring.	Champion Spring.	Red Spring.
Chloride of sodium........	390.127	459.903	141.872	509.968	400.444	562.080	458.299	506.630	702.239	83.530
Chloride of potassium.....	8.974	7.660	8.624	9.597	8.049	42.634	8.733	4.292	40.446	6.587
Bromide of sodium.........	0.731	0.987	0.844	1.534	8.559	2.212	1.307	0.266	3.579	Trace.
Iodide of sodium..........	0.086	0.071	0.047	0.198	0.138	0.248	0.039	0.906	0.234	Trace.
Fluoride of calcium.......	Trace.	Trace.	Trace.	Trace.	Trace.	Trace.	Trace.	Trace.	Trace.	Trace.
Bicarbonate of lithia.....	1.967	9.486	4.847	11.447	4.761	7.004	2.605	2.080	6.247	.242
Bicarbonate of soda.......	34.888	3.764	4.666	4.288	10.775	71.232	17.010	9.022	17.624	15.327
Bicarbonate of magnesia...	54.924	76.267	72.883	176.463	121.757	149.343	109.685	42.953	193.912	42.413
Bicarbonate of lime	131.739	120.169	93.119	170.646	143.399	170.392	96.703	109.656	227.070	101.256
Bicarbonate of strontia...	Trace.	Trace.	0.018	Trace.	0.425	Trace.	Trace.	0.082	Trace.
Bicarbonate of baryta.....	0.494	0.875	0.909	1.737	0.928	2.014	1.703	0.075	2.083	Trace.
Bicarbonate of iron.......	1.478	2.570	0.714	1.128	0.340	0.979	0.269	0.793	0.647	2.100
Sulphate of potassa.......	1.608	2.032	Trace.	Trace.	0.889	0.318	1.818	2.709	0.252	Trace.
Phosphate of soda.........	Trace.	0.007	0.016	0.006	0.016	Trace.	0.026	0.033	0.010	Trace.
Biborate of soda..........	Trace.	Trace.	Trace.	Trace.	Trace.	Trace.	Trace.	Trace.	Trace.	Trace.
Alumina...................	1.223	0.329	0.094	0.131	Trace.	Trace.	0.324	0.418	0.458	Trace.
Silica....................	2.260	3.155	3.184	1.260	0.840	0.665	2.653	1.145	0.609	3.255
Organic matter............	Trace.	Trace.	Trace.	Trace.	Trace.	Trace.	Trace.	Trace.	Trace.	Trace.
Total per United States gallon, 231 cubit inches..	630.500	687.275	331.837	888.403	700.895	991.546	701.174	680.436	1195.582	254.710
Carbonate acid gas........	409.458	332.458	245.734	375.747	392.289	454.082	384.969	344.669	465.468	....
Density...................	1.0092	1.0095	1.0035	1.0115	1.0096	1.0120				....
Temperature	52° F.				52° F.	46° F.	48° F.	48° F.	48° F.	....

cathartic leaves none of those unpleasant effects observable in the use of many other of the Saratoga waters. Adams & Jones are proprietors.

Having thus familiarized ourselves with the principal springs, we will, before noting other objects of interest, examine the appended analysis on opposite page.

Bottling the Waters.

The process of bottling is similar at all the springs, and as the Congress bottling-house is the most famous, a description of it will suffice for the whole, as given by Mr. C. C. Dawson of New York:—

"Probably not one-fifth part of the waters of these springs, which are used medicinally, are drank in Saratoga. Multitudes, it is true, flock here during the summer months; but their stay is usually limited to a few brief weeks—a time, in many cases, too short for these mild natural remedies to accomplish their perfect work. Thousands of visitors, therefore, find it necessary to continue the use of the waters after leaving the springs; and great numbers of other sufferers from the various ills which flesh is heir to, who are not able to visit Saratoga, still find the waters a source of comfort and health. Thus, while the benefit of these springs is enjoyed at Saratoga only by a comparatively limited number of persons, and principally during a brief season, their blessings are carried, by means of the bottled waters, all over the world, and are dispensed to multiplied thousands without regard to season or clime. A large and important branch of commerce has thus sprung into existence, involving a liberal expenditure of capital, and furnishing employment, directly or indirectly, to a great number of persons. The bottling and packing is carried on throughout the year; and, except during the height of the visiting season, when so much is consumed at the springs as materially to decrease the supply for bottling, the work is prosecuted night and day. The arrangements for this purpose are the most complete of anything of the kind in the country; and all the various operations are carried on with care, skill and perfection unsurpassed. In order to increase their facilities,

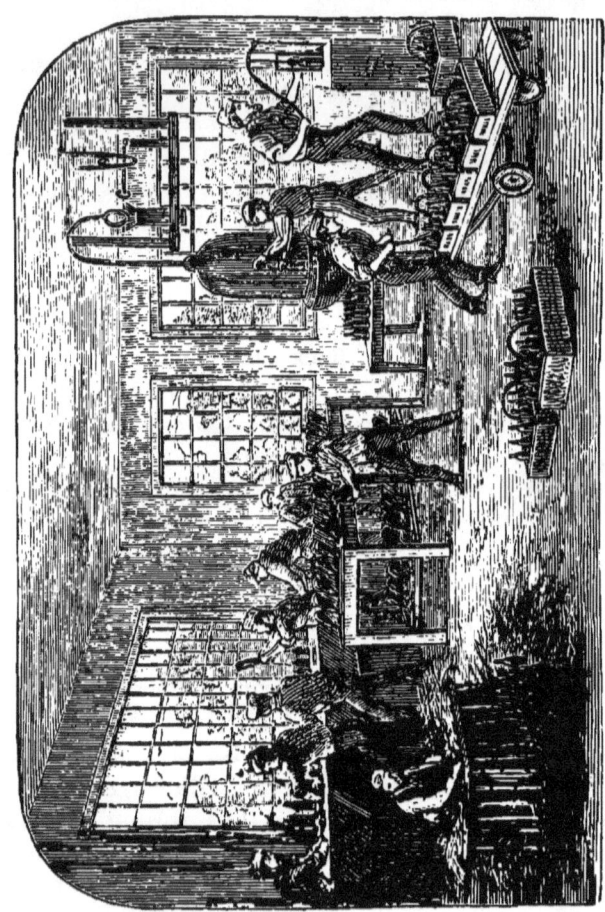

SCENE IN CONGRESS SPRING BOTTLING HOUSE, SARATOGA.
WASHING AND FILLING.

the Congress Spring Company have erected a glass factory near the village, where they not only make all the bottles required in their own immense business, but fill large orders for all kinds of bottles for other purposes. Some eighteen or twenty neat cottages in the same part of the village have been erected by the company for the use of their factory operatives.

"Each bottle, before being filled, is thoroughly washed and rinsed with both warm and cold water, a stream of each of which is constantly pouring into the tanks before the washers. To detach any impurities that cannot be removed by other means, a small brass chain is dropped into each bottle and thoroughly shaken about. The substitution of this simple and effective method of cleansing for the use of shot or pebble is an improvement which might well be adopted by every housewife.

"None but the finest corks are used; the brands used for branding them are set into a small table, their lettered faces being nearly level with its surface. They are kept hot by a jet of gas turned on them from below; and the corks receive their brand by being rolled over the heated types—an expert boy performing the simple operation with great rapidity.

"The water is pumped from the spring through pure block-tin pipes into a receiver holding from five to six gallons, from which it is drawn into the bottles; the pipes, pump and receiver being so constructed as to prevent any escape of the natural gases. The corks, after being soaked in warm water until they become so soft as to be easily compressed, are driven into the bottle by machinery, the process reducing their size before entering the bottles about one-third. It requires a strong bottle to stand the pressure of their expansion after being driven in; and even strong men sometimes find it difficult to pull them out. A single workman will fill and cork from fifteen to twenty dozen bottles per hour.

"After being filled and corked, the bottles are laid upon their sides in large bins holding from 150 to 200 dozen each, where they are allowed to remain four or five days, or longer, to test the strength of the bottles by the expansion of the gas, and also to detect any corks that may be leaky or otherwise imperfect. The breakage, while in this situation is about five per cent of the whole number filled, and sometimes more. The bottles frequently burst with a sharp report, like the firing of a pistol or the cracking of champagne bottles. Every bottle that breaks, either while in the testing-bins or in any of the various processes of washing, filling or packing, is registered in the office of the company by means of wires going from different parts of the establishment, and centring there in an apparatus arranged for the purpose All leaky corks are drawn, and the bottles refilled with water direct from the spring While all these precautions add. largely to the expense of putting up the waters, they render a leaky, and consequently a bad bottle of Saratoga water almost impossible; and they also render the breakage of bottles in subsequent handling a matter of rare occurrence.

"When the bottles and corks have been thus thoroughly tested, the corks are securely wired, this operation being performed with great rapidity by employees long trained to the work.

"The next process is the packing in cases, which is also done with great care and remarkable dexterity. The neck of each bottle is firmly wound with clean straw; and the bottles are placed on their sides in tiers of equal number, a parting strip of straw being laid between each bottle and its neighbor on either side. A layer of straw is also placed between the tiers of bottles, as well as the top and bottom of the box. When the box is filled, the packer walks over the bottles for the double purpose of settling them properly in their places, and as a further test of their strength, before the lid is put in its place and nailed down."

The Hotels of Saratoga.

Saratoga has long been famed for its grand hotels, wherein, during the season, the wealth and fashion of the whole country assemble, and where all the luxuries of a city home or the palace of a foreign nobleman can be found, within a minute's walk of the healing springs. Of all the great hostelries which have existed at the springs, the old United States has been one of the most noted, but that was burned several years ago. On its site, at a cost of over $1,000,000, has been erected and in 1874 opened to the public, the most stupendous, as well as the most elegant hotel in the country, if not in the world.

The United States Hotel.

This marvel of public houses is situated on Broadway and Division streets, extending on the latter fully back to the depot, with which it communicates by a lofty-arched passage for the convenience of guests who arrive in rainy weather. To say that the building is immense, conveys no idea of its proportions; that its internal arrangements are unequalled, will not describe them; that its park, promenades, ball-room, and wide verandas are beautiful, lovely, exquisite and delightful, cannot picture them as they deserve. The building covers seven acres of ground, and is arranged in the form of an irregular pentagon, having a frontage of 232 feet on Broadway, 503 on Division street, and 153 on Railroad place, extending back through all its length 54 feet. At the south end of the "main front" commences the "Cottage Wing," and extends back at right angles to the main building 566 feet. This wing is one of the prominent and peculiar features of the building, being intended to give families and parties the same quiet and seclusion which they could get in a private cottage, with the addition of the attention and conveniences belonging to a first-class hotel. In this wing the rooms are arranged in suites, containing from one to seven bedrooms, with parlor, bath-room, and water-closet attached to each suite. Here families can dine at their own table if they choose, and be in every way as much isolated as if in a private villa of their own. In the main front, on Broadway, is the grand drawing-room,

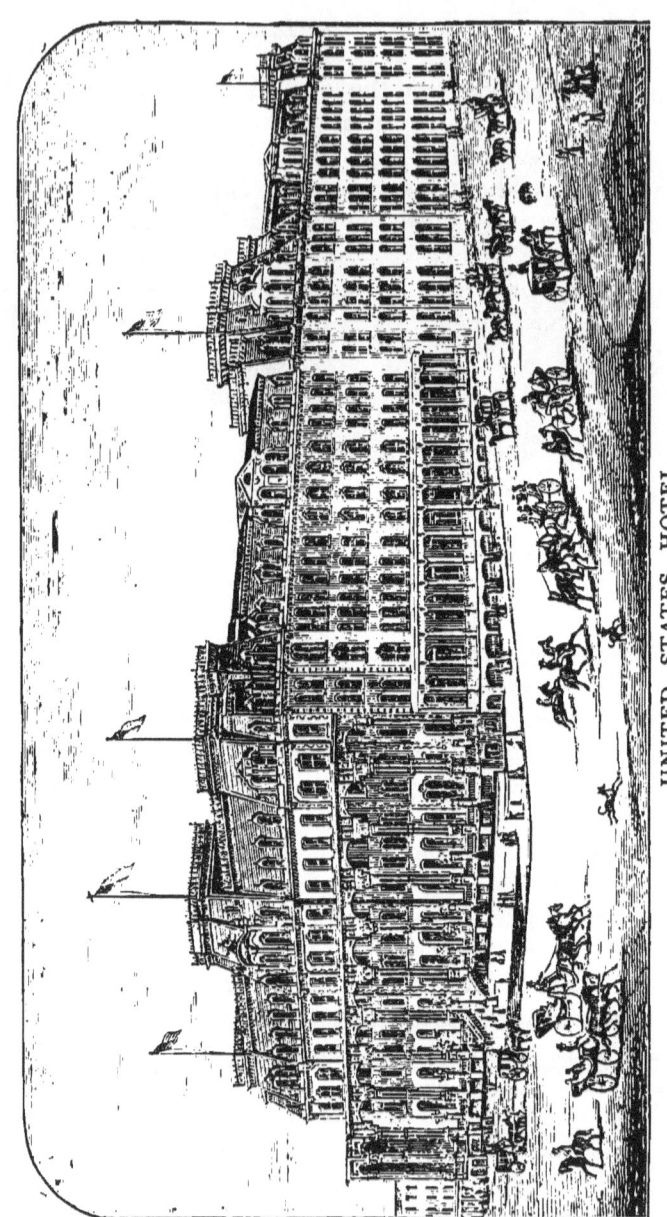

UNITED STATES HOTEL.

86 feet in length by 50 in breath, furnished in blue, with the finest Axminster carpets, carved black walnut and marble furniture, superb curtains and chandeliers. The chandeliers in this room cost each $1000.

North of the entrance hall is the ladies' parlor, furnished with exquisite taste. Next comes the gentlemen's reading-room, on the corner of Broadway and Division streets, connected with the business offices, in which is the largest telegraphic annunciator in the world, sixteen feet square. West of this is the grand dining-hall, fifty by two hundred and twelve feet, also the private dining-parlors, offices, wine-room, etc. The grand ballroom is situated in the second story of the Division-street wing, and is without doubt the finest room of its kind in the world. Connected with it is a quiet and secluded veranda overlooking the lawn. This retreat, dimly lighted, and secure from inquiring eyes as it is, where the strains of music from the ballroom are faintly heard, mingled with the plashing of the fountain beneath, and the murmur of the wind in the tree-tops which bend above it, is the most delightful spot imaginable for the exchange of those sweet nothings, which, far more than the music or the intoxication of the dreamy waltz, go to make up the fascination of the grand balls to susceptible young men and maidens. All the rooms throughout the house are furnished in the richest and most tasteful style, with running water in each. The elevators, two in number, are of the Otis Bros'. manufacture. One is intended solely for the use of arriving and the other for departing guests. The management is in the hands of the Hon. J. M. Marvin, the long-time proprietor of the old hotel. There are many other fine hotels in Saratoga; but they will not require a detailed description, for they have been long and favorably known to the travelling public. Each has some special point of attraction; but these have been so often enlarged upon, that the tourist feels perfectly acquainted with them. Prominent among them is

The Grand Union Hotel,

on Broadway, a short distance south of the United States, this season under the control of Messrs. J. H. Breslin, & Co. It accommodates twelve hundred guests. Across the street, directly opposite the Grand Union, is

Congress Hall.

This structure is four hundred and sixteen feet in length, and is supplied with every thing that can give pleasure, and add to the comfort of guests. At each extremity there are two large wings that extend back three hundred feet, and greatly enlarge the accommodations. Like the Grand Union, it occupies an entire square, covering all the space between Spring and Congress streets. Congress Hall occupies the site of a former house of the same name, which was burned in May, 1866. The proprietors are Hathorn & Southgate.

On Broadway, south of the Grand Union stood the Grand Central Hotel, which was leased by Mr. John B. Cozzens, but which was destroyed by fire in November, 1874.

On Broadway, South of the Grand Union, is the

Columbia Hotel,

which is owned and kept by D. A. Dodge, of Brooklyn New York. Its front forms one of the chain of hotels reaching from the west end of the United States to the Clarendon, and looks off on Congress Spring Park, one of the prettiest plateaus of Saratoga. This house is specially kept as a family hotel, homelike in apartments and moderate in charges, and will be found a pleasant stopping place.

Still south of the Grand Central, we come to another of the older Saratoga hotels, the Clarendon Hotel, Mr. Charles E. Leland, proprietor. This house is one of the most aristocratic at the Springs, and is too well known to require other than this passing mention.

The Holden House

is on Broadway, two doors north of the United States, and is owned and managed by Mr. C. H. Holden, who has just refurnished, recarpeted, repapered and repainted it throughout, and made it one of the handsomest as well as most comfortable of the smaller hotels of Saratoga. Mr. Holden has made many friends by his long connection with railroad and steamboat business, and the travelling public will naturally gravitate to his hotel. Everything is as complete in its way here, as at the large hotels, the bill of fare is good, the rooms neat and well furnished, and the prices are low. Mr. Holden is a gentleman whom his guests like, and carriages and porters meet every train on arrival in Saratoga.

The Waverley House.

Another of the small houses which deserves well of the public is the Waverley, on Broadway, a short distance north of the Town Hall. Major W. J. Riggs, its proprietor, is a genial, whole-souled man, and makes himself a favorite with his guests. This house accommodates 150 guests, without overcrowding, and is as well arranged as any of the large hotels in the place. The parlors and dining-rooms are large and airy, and furnished in excellent taste. The sleeping-rooms and private parlors are arranged in suites for the convenience of families and parties, or singly; and all communicate with the balconies, which extend around the house, and afford some of the loveliest views of the surrounding country to be obtained in this place. Among the chief attractions which this excellent house has for those who prefer health, quiet, and comfort to heat, dust, noise, and discomfort, are its large, airy, and well-arranged rooms. A further advantage is its moderate price.

There are a number of public institutions of various kinds in Saratoga, which are patronized by visitors, not only during the summer season, but also, to a greater or less extent, during the whole of the year. One of them is

Strong's Remedial Institute,

on Circular street, a short distance from Broadway and all the principal hotels and springs. Drs. S. S. and S. E. Strong, regular physicians, graduates of the University of New York,

STRONG'S REMEDIAL INSTITUTE.

are the proprietors. The institute was established several years ago, and has enjoyed a superior reputation for its treatment of invalids, as well as for its hotel and boarding accommodations. During the spring of 1871 the building was greatly enlarged, and now affords accommodations for 200 guests. Its parlors, dining-halls, and bath-rooms are fitted up in the most modern and elegant style; and the general appointments are of the first order. Being somewhat removed from the bustle and confusion of the larger hotels, it affords a delightful retreat for persons of impaired health; while refined and cultivated people will find its society more congenial than that of the more public houses. Among its annual patrons are the Rev. Theodore L. Cuyler, D. D.; Ex-Gov. Wells of Virginia; Mr. Robert Carter, of the firm of Carter Brothers, publishers, of New York; and many others of like position in society. The Institute is supplied with new and the most improved appliances now known to medical science, among which are the Electrothermal, Sulphur, Air, Turkish, and Russian Baths; Swedish movement cure; the Equalizer or Vacuum Treatment; Oxygen Inhalations; Gymnastics; and other varieties of hydropathy and medicine.

Temple Grove Seminary is beautifully situated in a grove, on what was formerly called Temple Hill; and its grounds occupy the whole square on Spring street, between Circular and Regent streets.

Social Life at the Springs.

But the natural advantages alone of Saratoga would have never given it the pre-eminence above all other wateringplaces which it enjoys. The results of human art and the enjoyments of social and fashionable life are the chief claims upon the favor of many visitors. Not even the invalids come here to drink the waters alone; they expect to meet and enjoy the society of other invalids, and the gayeties of the season. Dancing and drinking are reckoned by some as the chief employments of guests at Saratoga, and so far as the morning draught of two or half a dozen glasses of spring water, and the nightly hop at one of the—

> "—— great hotels ablaze with light,
> Where youth and beauty, wealth and rank,
> Hold revel through the night,"

are concerned, the truth justifies the declaration. But there are other and even more entrancing social pleasures at the Springs, of which we will enumerate a few. First, there is the Lake Drive. Saratoga Lake, a beautiful sheet of water nine miles long and five miles wide, is four miles distant from the village by the extension of Union Avenue. The drive is a continuation of East Congress Street, and has a row of trees each side and one in the middle. A most gay and brilliant scene is presented on a bright August morning or afternoon, as the long procession of carriages, in all the richest styles, pass down on one side of the drive and back on the other. On a high bluff, near the outlet of the Lake, is Moon's Lake House, kept for the accommodation of the many visitors who every fine day ride down from the Springs. A mile beyond the Lake House is Chapman's Hill, which rises 180 feet above the surface of the lake; and three miles farther on is Wagner's Hill, 240 feet high. N. P. Willis relates, among his legends, the following tradition of Saratoga Lake: "'There is,' he says, ''an Indian superstition attached to this lake, which probably has its source in its remarkable loneliness and tranquility. The Mohawks believed that its stillness was sacred to the Great Spirit, and that, if a human voice uttered a sound upon its waters, the canoe of the offender would instantly sink. A story is told of an Englishwoman, in the early days of the first settlers, who had occasion to cross this lake with a party of Indians, who, before embarking, warned her most impressively of the spell. It was a silent, breathless day, and the canoe shot over the smooth surface of the lake like an arrow. About a mile from the shore, near the centre of the lake, the woman willing to convince the savages of the weakness of their superstition, uttered a loud cry. The countenances of the Indians fell instantly to the deepest gloom. After a moment's pause, however, they redoubled their exertions, and in frowning silence drove the light bark like an arrow over the waters They reached the shore in safety and drew up the canoe, and the woman rallied the chief on his credulity. 'The Great Spirit is merciful,' answered the scornful Mohawk; 'he knows that a white woman cannot hold her tongue.'" Whatever basis there may be for this legend, certain it is that the Great Spirit has removed the prohibition, if any existed, upon talking and laughter upon the lake, as witness any pleasant evening, when parties who have ridden out to Moon's are enjoying the delight of a moonlight sail or row." The collegiate regattas of 1874 and 1875 on this lake have attracted hosts of visitors hither, and have drawn general attention to the merits of this beautiful sheet of water as a race course for shells.

Then, in the season, there are the races. The Saratoga race course is only a mile from Broadway, near Union Avenue, and it is always kept in fine condition. The attendance at the races embraces a large share of the wealth and style of the country, and the grand stand is filled not only with interested turfmen, owners of fleet horses and gentlemen of means, but with hundreds of stylish and elegantly dressed ladies, who appear to be as much excited over the contests as their male neighbors, and who freely wager such trifles as a dozen gloves or a bottle of wine on their favorites. Then there are the weekly balls at the principal hotels. At these, of course, dancing is subordinated to dress. All the ladies from each of the other hotels will, of course, make it a point to stroll over in the course of the evening and see what everybody "has on," and it is of course necessary to dress so that their feelings will be inexpressible. Hence the size and number of the Saratoga trunks which every lady finds a necessity of her outfit for the Springs. Hence, also, the reports in the Saratoga and the metropolitan papers of the dresses of the belles. Many of these glowing descriptions are inserted by special request. When we read that "Miss Clementina Van Tassel was much admired in a *jupon* of mauve *matelasse*, with *peignoir* of satin *tarletane* cut bias and ruffled to the waist with pink *tulle;* corsage bouffant and panier *decollete* with point lace *en traine, a la maitre de hotel;* hair *a la cuspidore* dressed with callas and nasturtiums, (the description may not be just right in some of its details, but if so any of our lady readers will correct us) we take it for granted that she or her escort has given the details to the society reporter, who is always glad to spice his description of the ball with these personal " puffs." The scene in the grand ball-room of the United States, Grand Union, or Congress Hall, on one of the ball nights fully justifies this glowing verse of Miles O'Reilley's:—

> A fairy scene of colored light,
>   Of gorgeous dress and magic changes,
> Where still the gazer's dazzled sight
>   From beauty to new beauty ranges.
> Now rings the music clear and high,
>   Now seems to die; now swells in clangors;
> Voluptuous visions fill the eye
>   And thrill the pulse with tropic languors.

Flirtation and its Concomitants.

And last, but most fascinating of all, there is — flirtation. Without this chiefest of watering place charms, social life even at Saratoga would lose its attraction. The promenade on the balcony between dances; the whispered word or the sly glance over the morning glass at the spring; the moonlight drive to the Lake, the "holding hands" on the Clarendon piazza or on the benches in the Park — all these delights would be blotted out and Saratoga would become to the young and impressible, a dreary waste, a howling wilderness — that is, if a wilderness ever howls. J. Cheever Goodwin, the talented young poet and dramatist of Boston, has illustrated in graceful verse the romance of a Saratoga flirtation, as follows : —

It was up at Saratoga that I met her,
　Where I went to drink the waters for my health ;
And her stylish way (I never shall forget her)
　Seemed to me a sure concomitant of wealth.

In her figure and her face she was a Venus ;
　Like the evanescent lightning shone her eyes :
In the dining-room one table was between us ;
　But love such paltry distances defies.

I smiled my adoration o'er my coffee,
　Drank deep of tender passion with my tea :
As the waiter took my trout untasted off, he
　Little thought it was so typical of me.

I was caught as fast as ever were the fishes ;
　And the hook went deeper in with every meal :
But my hopes were all as empty as the dishes ;
　And my sorrow cut as deep as knife of steel.

'Twas in vain I promenaded the piazza :
　She was never in the parlor night or day ;
And I thought, " She is an invalid, and has a-
　N injunction in her room to always stay.

" For I never find her drinking at the Hathorn ;
　To the hops or balls I never see her go ;
She is never betting Belmont or McGrath on,
　At the races where so many beauties show."

My suspicions were, alas ! substantiated ;
　For excepting at our meals we never met :
You 'd have surely thought I was a man she hated,
　Excepting for the smiles I used to get.

" Does she ever think of me? " I sadly wonder :
　" When she 's seated at her breakfast or her tea,
Through the many miles that keep us now asunder,
　Does her memory ever send a thought to me? "

And I sadly fear I 'm utterly forgotten,
　That my presence would not cause her heart to stir,
That she 'd give to see me not a single button,
　Though I 'd gladly give a dozen to see her.

Schroon Lake and Neighboring Resorts.

Taking Saratoga as our point of departure, there are several excursions which we can make with ease, and by which we can visit many scenes of interest in a very few days and by comparatively very few miles of travel. The Adirondack Railroad, which is pushing its way northward from Saratoga into the heart of the great wilderness only trodden by the wild deer and the Indian until within a few years, affords easy communication with a number of noted resorts, embracing views of Nature in all her pristine beauty and savage grandeur, within a few hours' ride of the grand hotels of the Springs. A trip of fifty miles in the cars brings us to Riverside station, but a few miles short of the present terminus of the railroad (North Creek), whence we take a stage for a six mile ride to the foot of Schroon Lake, a favorite and very beautiful resort. Schroon Lake is unique as a watering-place, being in a semi-wilderness, — the outskirts of the Adirondack region, — and surrounded by wild and rugged hills, yet as has been shown, within a few miles of the railroad, which is in effect the same as if it were within twenty miles of Saratoga by stage. Accordingly, we shall be very apt to find the cars well filled and outside seats on the stages in lively demand, if we go to Schroon in the Summer season. The Lake is nine miles long, and averages about two miles wide, except at the Narrows, which contract to half a mile or less in width, about midway its length. At the upper end is the village of Schroon Lake; at the lower end, where the stage deposits us, is Pottersville. The outlet is at the latter point, and the water finds its way through Schroon river to the Hudson, near Warrensburgh, 20 miles south. Schroon Lake is also reached by staging from Lake George to Thurman, nine miles thence by rail, 14 miles to Riverside. At Pottersville, the hotel kept by L. R. Locke, affords us a good dinner or accommodations for a longer stay if we choose, teams for a trip into the wilderness or conveyance to the Landing at the foot of the Lake, where the little steamer Effingham is in waiting to convey us to Schroon, through the entire length of the beautiful sheet of water. A little to the north of the village a swift mountain stream, on its way to the Lake, has worn a channel through the white marble, which forms a natural arch above it, 40 feet high and 247 feet long. The lake, which is really a widening of the Schroon River, or northeast branch of the Hudson, is at a level of 1000 feet above tide water, and has but a single island, Isola Bella, near the northern extremity. There have been various attempts to assign an Indian paternity to the name Schroon, but it is far more likely that the tradition which assigns its name to the early French occupants of Crown Point, who gave it the title of "Scaron," after Madame de Maintenon (Scaron), the second wife of Louis XIV. From the landing, the little steamer shoots across the lake northeasterly, to a considerable bay on the east shore at the head of which Mill Brook empties into the Lake. Near its mouth, a new hotel, the Wells House, is attracting a good share of patronage. Thence in a long, sweeping curve, we plow through the Narrows and lay a straight course for the little cluster of houses interspersed with hotels which form the village of Schroon, passing on the right Isola Bella, with its villa and gardens. Here is the Leland House, overlooking half the lake from its elevated site on a projection from the west shore; near by the Wickham House, new and commodious, and the Taylor House and Ondawa House to the left as the boat stands in. At any of these hotels, comfortable rooms, the best of mountain fare, boats and guides can be had at reasonable rates, and the visitor can highly enjoy a day's or a month's stay, varied by excursions down the Lake or into the woods; the ascent of Mount Pharaoh and Mount Severn, and the fishing in Pharaoh Lake at the foot of the mountain of the same name, whence trout are taken in great numbers. Paradox Lake, nine miles north, is much visited. It is a lovely and secluded pond, with romantic scenery and good fishing for its recommendations It is one of the feeders of Schroon river, and it derives its name from the fact that it is so little above the level of the stream, that in the Spring freshets the water of the river runs into it, instead of out. Long Pond, Pyramid Pond and other small and nameless little tarns lie in its immediate vicinity, Pratt's hotel, near the head of Paradox, being the head-quarters of visitors to the region.

Returning to Pottersville by steamer, and thence by stage to Riverside, we may find it worth while to stop at Charlestown, six miles from Pottersville, and thence make discursions to Brant Lake, nestled among the Kayaderosseras peaks, in the town of Horicon, and the other points in the vicinity. Returning by rail from Riverside to Hadley, 28 miles, many tourists alight for a visit to the village and Lake of Luzerne on the opposite, or east bank of the Hudson. The lake is a quiet and picturesque little body of water, among the hills, 700 feet above sea level, with a single island. In the village are the Wayside Hotel, Rockwell's and the Wilcox House, all of good repute, and here boats can be procured for the navigation of the lake, and teams for the many fine drives in the vicinity. The following lines by Percival find an echo in the feeling of visitors to this lovely little mountain mirror:

> The waves along thy pebbly shore,
>   As blows the north wind, heave their foam,
> And curl around the dashing oar,
>   As late the boatman hies him home.
>
> How sweet at set of sun, to view
>   Thy golden mirror, spreading wide;
> And see the mist of mantling blue
>   Float round the distant mountain side.
>
> At midnight hour, as shines the moon,
>   A sheet of silver spreads below;
> And swift she cuts, at highest noon,
>   Light clouds like wreaths of purest snow.
>
> On thy fair bosom, silver lake.
>   O, I could ever sweep the oar;
> When early birds at morning wake,
>   And evening tells us toils are o'er.

The village of Hadley stands at the junction of the Sacondaga river with the Hudson. The Indians called it Tiosawonda, or the "Meeting of the Waters." Near by are Jessup's Little Falls, or Luzerne Rapids, where the Hudson dashes between lofty banks over a declivity of 18 feet. Jessup's Great Fall, five miles below, is 70 feet high, and is much visited. From Hadley we cross the Sacondaga river on a bridge 450 feet long and 96 feet above the bed of the river.

The Route to Lake George.

But by far the greater proportion of visitors to Saratoga make their principal excursion thence to Lake George. This far-famed lake, which has no rival in this country so far, for gradeur and beauty of scenery, facility of access and superb accommodations, is usually visited by thousands of tourists, while many wealthy families from New York and other cities own islands or seats on the shores where they spend a portion of each summer. Many of the villas are of great elegance, and the air of the entire locality and its visitors is that of refinement and luxury. The lake is about thirty-one miles at its nearest point, Caldwell, from Saratoga, and is reached by taking the cars of the Rensselaer and Saratoga Railroad to Fort Edward, seventeen miles northeast, and by the Glen's Falls Branch, five miles northwest, to Glen's Falls, on the Hudson river, two hundred miles from its mouth. This village, which has a population of about 8,000, is one the of most thriving and certainly one of the handsomest of inland towns. The Rockwell House, fronting on the fine public square — with its Soldiers' Monument surrounded by an eagle and flanked by military statues; its lofty and handsome fountain, which, perpetually playing, scatters its misty spray over the broad street and delightfully cools the air; and with its tree-shaded and neatly kept streets diverging in various directions — is a fine, commodious house, much frequented. The industries of Glen's Falls comprise its quarries of black marble and limestone; its paper and saw mills and its trade with the country around. Its chief feature of interest is the great fall of the Hudson from which its name is derived. The substratum of this region is black limestone, which is crystallized in places, and so regularly stratified that a perpendicular section looks like hewn stones in the wall of a building. The action of the water has worn some of these strata away, a few at the top, and more farther down the falls; so that a kind of irregular series of steps has been formed, over which the waters of the river go thundering down a descent of over fifty feet. Seen in the sunlight, rainbows appear in the clouds of spray that are tossed into the air just below. The river has worn its way deep into the black limestone, which rises in some places to the height of seventy feet above the surface of the river. A bridge six hundred feet long, which rests on a marble island in the centre, crosses the Hudson at this point, and from it one of the best views of the falls is obtained. By a private stairway that goes down near the bridge, one may reach two objects of interest, Indian Cave and Big Snake. The cave runs through a small island, from one channel to another. This is said to be the place of concealment of Cora and Alice, Mayor Hayward and the singing-master, characters familiar to the readers of Cooper's "Last of the Mohicans." Here David blew his pitch-pipe and sang "The Isle of Wight" to the accompaniment of the roaring waters, and here Uncas watched over the slumbers of the fair sisters. It is a capital place to grow romantic, and after a scramble through the rocky cavern, one can imagine a Mingo hidden in every bush on the shore, and can hear the scalp-song in every note of the rushing stream. "Big Snake" is the name applied to the likeness of a serpent in a vein of stone projecting from the smooth surface of a softer ledge. In the rocks near the falls many trilobites and others of our fossil ancestors (according to Darwin) are entombed, but no one seems to weep over the tomb of his remote progenitor. The Darwinian theory is all very nice — for other people — but somehow, no one wants to take it home to himself. Caldwell, at the head of Lake George, is nine miles from Glen's Falls, which is reached by a romantic and view-affording stage ride by the plank road. As we rattle out of Glen's Falls in fine style, occupying, if the day be fair and we lucky, seats "on top" of the Concord coach, drawn by its four spanking roadsters, through the toll-gate, past the fair grounds and then over hill and dale in a nearly direct northeasterly course — by ponds starred with fragrant pond lilies, through dense woods whose arching branches unite over our heads and sometimes almost scrape us off the coach as we dash beneath them — to the "Half Way House," where George Brown, the jovial host, is always ready to get up a milk punch or a lemonade for each thirsty soul. As we traverse the region of the ponds we are waylaid by Bedouins, that is, we suppose they are Bedouins, young Arabs laden with the white lilies tied up in odorous bunches, which they shy into the coach windows, and with which they bombard the roof passengers, meantime trotting alongside the team in expectation of "back-shish" which they generally get in the form of a hail-storm of coppers. Shortly after passing the Half Way House we see the Williams Monument, a marble obelisk eight feet high, standing on a huge boulder upon a side hill to the left of the road. The inscription reads: —

ERECTED TO THE
MEMORY OF
COLONEL EPHRAIM WILLIAMS,
A Native of Newton, Mass.,
who after
GALLANTLY DEFENDING
the frontiers of His Native State,
SERVED UNDER GEN. JOHNSON
against the
FRENCH AND INDIANS,
and Nobly Fell Near this Spot
in the Bloody Conflict of
September 8, 1755,
in the 42d year
of His Age.

The monument tells its own story, and it is only necessary to add that Colonel Williams was the founder of the college which bears his name at Williamstown, Mass., and that the monument was erected by the students in 1854, to perpetuate the memory of a good man and a brave soldier. All this section of country has been the theatre of conflict from the first white occupation down to the close of the Revolution, and soon we come in sight of a lovely little pond, now shaded by bending trees, and dotted with snowy lilies, which bears the strangely inappropriate name, as it now seems, of Bloody Pond. Its name is derived from one of those sanguinary surprises of the "old French war," when a party of the French, cooking heir supper around this little lakelet, were ambuscaded by the English and slaughtered in such numbers that their blood is said to have tinged the water red. Soon after, we gain our first views of the "Horicon Water," gleaming through the trees like a gem of lapis lazuli, set in the emerald of the foliage. From this point the road winds around the hillside and down to the lake, and we see the white houses of Caldwell almost at our feet, and "catch the gleam of a passing sail" or the puff of fleecy smoke from one of the little steamers that ply on the lake. In a few moments we turn up a gravelled driveway and alight upon the long piazza of the Fort William Henry Hotel. Lake George is before us, and the mountains behind and about us.

CHAPTER V.

Lake George and its Beauties.

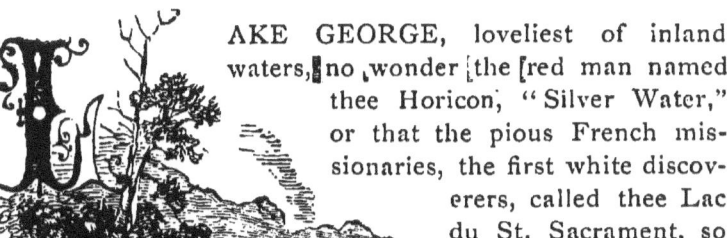

AKE GEORGE, loveliest of inland waters, no wonder the red man named thee Horicon, "Silver Water," or that the pious French missionaries, the first white discoverers, called thee Lac du St. Sacrament, so pure, so clear, so brilliant are thy waters! Of all the liquid jewels shining on the earth's brow, thou art the fairest and the brightest! Romance, tradition, and history, combine to invest thee with a poetic charm! Hither for yeares th tide of summer travel has tended, and here the pilgrim, seeking what is fairest and sweetest in Nature, is content to abide, sure that nothing can surpass the beauties around him. Lake

George lies in a valley whose lower outlet is far at the north, whose walls are mountains, and whose charms have been the inspiration of poets and painters for a century. This lovely body of water lies about 300 feet above the sea level, and stretches thirty-five miles in length, from north to south, with a varying width of from one to four miles. It is supplied by springs and mountain brooks, and hence its waters are clear as crystal and cold as the snows of Lebanon. Its shores are dotted with the villas of wealthy summer residents, and its surface studded with verdant islets, most of which are crowned with tasteful and romantic cottages. Caldwell, at its head, or southern extremity, is a little hamlet whose chief industry is the reception and accommodation of summer visitors, and which is largely composed of hotels, the Fort William Henry, the Lake House, the Harris and the Central, while just across the Lake, embowered in trees, stands Crosbyside, on a beautiful slope — a quiet and unexceptionable retreat. The Fort William Henry Hotel, from whose broad piazzas and tasteful grounds a splendid view of the watery expanse is gained, is one of the most famous, most elegant and most fashionable of watering place resorts. It is from four to six stories high, crowned with a Mansard roof and two lofty Renaissance domes, and fronts 334 feet on the lake side. A piazza twenty-five feet wide, and shaded by a roof supported on columns thirty feet high, extends along the entire front, and here of a summer's day or evening, a bewildering array of dainty feminine toilettes and astonishing masculine finery is spread out to view. Of course, the animated contents of the garments aforesaid are but a secondary consideration, but if one cares to examine closely, he will find that they comprise in no small proportion the elite of metropolitan society. In the evening the blaze of light from the halls, parlors and the central office, is enough to dazzle one; but the gas jets are eclipsed in splendor by the flashing radiance of the diamonds worn by the lady guests. Strains of delicious music float out through the open windows, and within, the whirl of lithe forms and the shimmering of silks and fleecy fabrics in the dreamy waltz makes up a scene of gayety which is hardly matched even at Saratoga. In the grand dining hall, brilliant with table ware, glass and silver, the dusky cohorts wheel and deploy in never ceasing action, obedient to the wish of the guest and the rules of that mighty potentate, the head waiter, seemingly engaged in a never-ending endeavor to supply a never-satisfied demand for "Lake George trout, with fried potatoes and coffee." Above, are rooms for 900 guests, and rarely, during the season, are the accommodations sufficient for the crowd of visitors. From the front of the house, elegant and tastefully ornamented grounds slope to the water, with a plashing fountain in the centre, making music in the little basin, on which snowy swans float to and fro. Broad gravelled walks radiate in every direction, and the dark shrubbery is lit up at night by the radiance of gas lamps conveniently disposed. The ruling spirit of this great caravanserai is Mr. Roessle, who with his sons have been the proprietors since 1868, and who has made his name synonymous with the highest luxury and elegance of which a hotel is capable. There are many points of interest in the vicinity; visits to the ruins of Forts Williams Henry and George, the ride to the Healing Springs, and that up Prospect Mountain being the favorite trips; but the enjoyment best of all is found in the sail down the Lake.

A Trip Down the Lake.

From the landing at the water-side, the little steamer Minnehaha steams daily to the foot of the lake and back, and many tourists avail themselves of the opportunity to spend a delightful day in the round trip. Swinging out from the little wharf, the vessel glides out into the lake, with French Mountain looming grandly on the right, and Rattlesnake Cobble frowning behind us. We pass Tea Island, about a mile from our starting point, a small, wooded islet near the west shore, with a rustic cottage, and memories of a "tea-house" kept there years ago for the regalement of visitors. Diamond Island, a mile and a half further, lies on our right hand, and derived its name from the quantities of quartz crytals formerly found here. In 1777, Burgoyne fortified it as a military post, and the same year it was the scene of an encounter with the Americans. A mile further, on the left bank, stands the Coolidge House, a quiet, pleasant resort, much sought by fishermen, who find in its vicinity some of the finest fishing on the lake. Nearly opposite, out in the lake, lie the Three Sisters, and still further east Long Island, the largest in Lake George, extending about a mile North and South, and affording room for a fine farm on its surface. A steamboat dock extends from the east side, and after touching here, we round the northern end of the island, and bear westerly to Bolton. On our right, the lofty, rugged Buck Mountain towers 2000 feet above us, and just south of it is Pilot Mountain, only less lofty. Before reaching Bolton, we pass Dome Island, the loftiest and one of the largest in the lake, oval in form, and sufficiently dome-shaped to justify its name. We next pass Recluse Island, owned by Rufus Wattles of New York, and one of the loveliest gems on the bosom of the lake. It is handsomely wooded, and bears a tasteful cottage, with outlying pavilions, summer houses, and a graceful bridge connects it with the tiny Sloop Island on the east. Recluse Island became famous in 1868 by a newspaper hoax, which represented it sunk by an earthquake—a statement which proved " not sufficiently materialized."

Bolton, ten miles from Caldwell, is the second village in size upon Lake George, and possesses three excellent hotels, the Mohican, the Bolton and the Wells Houses, all chiefly patronized by season boarders. Its site is deeply embayed, and its view down the lake includes the Narrows, formed by Tongue Mountain on the left and Shelving Point on the right; and Northwest Bay, a considerable indentation of the west shore formed by the projection of Tongue Mountain. Bolton presents a most romantic appearance from the water, the many beautiful cottages, the shaded grounds, and the pretty little church of St. Sacrament, built by the exertions of a young daughter of Mr. Thieviot, giving it the aspect of one of those picturesque French villages which painters delight to put on canvass. Tongue Mountain to the north, already mentioned, is a beautiful and shapely elevation, whose contour indicates the derivation of its name. Green Island, near the Bolton shore, is about half a mile long, and is a beautiful spot. and Crown Island near by is one of the gems of nature. From Bolton our course is nearly east, to Fourteen Mile Island, at the extreme point of Shelving Rock, which forms the right wall of the Narrows. These Narrows are among the most lovely of the features of Lake George. Here the shores are less than a mile apart, and so far do the lofty banks overlap each other, and so full is the course of islands, that from the steamer it looks as if the lake ended here. But we touch at the little pier of Fourteen Mile Island, 12 miles from Caldwell, and a favorite resort for artists and gentlemen of a piscatorial turn of mind, and swinging out again see clear water beyond. The view through the Narrows during the changing moods of a summer afternoon, and especially during a summer shower, is inexpressibly grand. The lofty peaks on either hand clothed in purple mists, the rolling vapors which curl around their rugged sides, the drifting clouds which now and then eclipse their summits, and above all, the sunshine breaking through the rifts and glorifying the scene—all make up a picture which would drive an artist frantic with the knowledge of his inability to reproduce it on canvass. Laying our course nearly due north from Fourteen Mile Island, we are soon passing Black Mountain, the highest on the lake, being nearly 3000 feet in height. Its sides are clothed in dense woods, two-thirds its height, and above, the bare-bleak rocks tower into the clouds. It is often ascended with the aid of guides, and from its summit in a clear day, views of the entire lake and the surrounding country are gained. Next beyond is Sugar Loaf, a spur of Black Mountain and a rude, lofty mass of rock. In the lake are the Harbor Islands, the scene of a fierce battle in 1757, in which some 300 or 400 English were worsted by a band of Montcalm's Indian allies. Deer's Leap Mountain is on the west bank, and gains its name from the fact that a deer, pursued by hunters and hounds "took a header" off the beetling precipice, and was impaled on a sharp tree top below. On the right shore, in a smooth hollow, sloping down to the beautiful Bosom Bay, nestles the little hamlet of Dresden, which is reached from Hulett's Landing on the point south of the bay. Again striking across the lake to the west shore, we pass Sabbath Day Point, a bold projection, tradition says named because Abercrombie halted his troops here over Sunday, on their way to attack Ticonderoga. Prosaic history destroys the romance of this story by insisting that Abercrombie landed here on a Wednesday, and leaves the origin of the name still in the dark. Bluff Head is the point on the right shore, opposite Sabbath Day Point, and the long ridge back of it is called Spruce Mountain. Six miles north of Sabbath Day Point, and twenty-eight from Caldwell, at the head of a semicircular bay, lies the village of Hague, with its Phœnix Hotel and its fine fishing grounds in the broad lake. Just above is Friend's Point, off which lies Waltonian Isle, named from a party which formerly camped there. On the east shore, thirty miles from Caldwell, is a huge projecting hill, known as Anthony's Nose, off which the water is said to be the deepest of any in Lake George. Two miles further down, on the west shore, is Rogers's Slide, a smooth, nearly perpendicular wall of rock, 400 feet high, whose base is in the lake bottom. The story whence its name is derived is that in the winter of 1758 Major Rogers, who commanded a company of Colonial soldiers, was scouting near the outlet of the lake, and was discovered and pursued by the Indians. He came to the high bluff, near the summit of the slide, and made his way down to the upper edge of the inclined plane; here he unfastened his snow-shoes, turned about in them, and with his toes towards the heels of his shoes, walked away from the rock, took a circuit down to the ice, and made his escape to Fort George. The Indians came to the top of the rock, and, seeing apparently the tracks of two persons directed towards the lake, they supposed that two men must have slid down the rock; this belief was strengthened by the sight of the major running across the ice. The Indians were filled with wonder that any man could go down this long and steep descent, and find himself alive afterward; and they felt sure the major must have been under the protection of the Great Spirit, and dared not further molest one who had defied such danger. There seems to be no authority for the story, but Rogers was noted for his "Munchausen yarns," and very likely he told it. From this point the banks grow lower and less picturesque, the water shoals, and we approach the foot of the lake. Prisoner's Island, near the west shore, is so called from a tradition that in 1758, Abercrombie confined a lot of French prisoners on this island, whence they escaped by wading ashore. But while we are pondering on this remarkable statement, the boat nears the dock, a few bumps, a grating against the timbers, and we are fast to the pier, near which the cars are waiting to convey us to Lake Champlain.

The Geology and History of Lake George.

Where Lake George now reposes was once a valley, bounded by low hills of the primitive limestone rock, but the "drift period" of the geologists flooded the valley with a mighty deluge, and covered hill and dale with gravel, sand and soil. The flood passed, and the lake fell to its present level, the islands and hilly banks emerging clothed with soil, and the bed of the lake being covered with snow white sand. The water is of remarkable purity, so that objects can be seen at a great depth. Travellers liken Lake George to Loch Katrine in Scotland. The region around the lake, has been harried by the movements of hostile armies, until nearly every point bears some historic or legendary repute. Even the islands were the home of brave and daring rangers, and some of them were even fortified, attacked, defended and captured, as military posts. Caldwell stands near the site of Fort William Henry, which was erected by Gen. Johnson in 1755, after the battle near Bloody Pond already referred to. It was at this fort, in 1757, that the Indian allies of the French marquis, Montcalm, fell upon the English, who had surrendered themselves to the French as prisoners of war, and murdered in cold blood or carried away captive fifteen hundred men. The ruins of Fort George are about a mile south-east from the Fort William Henry Hotel. All that is now left of the old fort is the ruins of the rectangular citadel that was built inside of the breastworks. A part of the old wall, nearly twenty feet high is standing.

In 1609, Champlain ascended the St. Lawrence and crossed the lake with which his name is ever since associated, to a point near the present site of Ticonderoga. His Indian companions described Lake George to him but he never entered it. The first white man probably who saw the lake, was Father Jogues, the Jesuit missionary, who in May, 1846, attended by Jean Bourdon, the engineer, of Quebec, arrived at the outlet of the lake, on the eve of the festival of Corpus Christi, and in honor of the day gave the water the name of Lac du Sacrement, or Lake of the Blessed Sacrament. During the twenty years succeeding, visits were made to Lake St. Sacrament by the French from Canada, and in 1691 Major John Schuyler left Albany with a force which scouted up and down the lake. During Queen Anne's war, from 1702 to 1713, the lake was used as the route to Canada, as was also the case during the war of 1745—1760, when the peaceful quiet of its woods and waters was often broken by the rude shock of battle. In 1755 the battles near Williams Rock, in which Colonel Ephraim Williams and the old Mohawk Chief Hendrick were killed; the repulse of the French from Johnson's camp, near the subsequent site of Fort George, and the rout and destruction of the French force at Bloody Pond followed in rapid succession, and during the entire war, Lake George was the scene of carnage and disorder. The story of the siege of Fort William Henry and of the various battles on the lake, can be found in any history of this country. About 1770, peace having been restored, settlers began to locate on the shores of Lake George, and Forts George and Ticonderoga were substantially deserted and fast going to decay. In 1775 the Revolutionary excitement began to be felt about Lake George, and the capture of Ticonderoga by Ethan Allen gave the possession of the whole section into the hands of the colonists. With these events the military history of the lake ceases.

Ticonderoga and the Road to Lake Champlain.

But while we have thus been reminiscently engaged, the train has been waiting for us on the dock beside us, to convey us to Lake Champlain. Up to the fall of 1874, the trip was made on Baldwin's stages, and one of the events of the journey was the oration delivered by the proprietor, who always accompanied them, on reaching the ruins of old Fort Ti. But now, a branch railroad from the New York and Canada, which is slowly eating its way through the lime-stone cliffs on the west shore of Lake Champlain, and which passes by a tunnel underneath the promontory on which stand the ruins of the old fort, conveys us in a few moments from the landing, a few rods above the old dock at the foot of Lake George, along the outlet to the shore of Lake Champlain, beneath Mount Defiance.

The outlet of Lake George is by a small river or large creek, which describes almost a horse shoe curve, and falls 240 feet in its course of four miles. The Indians called it Cheonderoga or "Sounding Water," from its perpetual music, and the present name Ticonderoga is but a colonial corruption

RUINS OF "OLD FORT TI."

thereof. The French settlers called it Carillon, meaning a chime, from the same cause. The course of the creek is broken by two considerable falls, and by almost continuous rapids. Our road follows the left bank of the stream quite closely for a few rods, then crosses the creek and passes through the little village of Ticonderoga which lies near the lower falls. Then we continue along the stream, round by a sweeping curve to the right the base of the hill, which expands into the lofty eminence known as Mount Defiance, and bring up at the dock on Lake Champlain. Just across the bay into which the "outlet" widens, stands the Ticonderoga of history. We may if we choose, ascend Mount Defiance, and from its summit gain a wide and interesting view of the points so re. nowned in the old wars which were waged in this vicinity.

Historical and Descriptive.

Ticonderoga as we have seen, is an elevated point of land, with water on three sides, and is well adapted for defense. The first fortification was built in 1691 by Colonel Philip Schuyler, when he was on his way to attack the French at Laprairie. In 1755, the French works begun 25 years before at Crown Point being still weak and insufficient, Montcalm decided to build a new fort at Carillon, and up to 1759 the construction of works on this peninsula was actively pushed by the French. In the last named year they were evacuated and partially blown up, and Colonel Eyre planned a new work, but it was never completed, and in 1773 the fort was in a ruinous condition. May 10, 1775, Ethan Allen, Benedict Arnold, and 85 Vermont and Massachusetts men surprised and captured the fort, which they held till July. In that month, the British having erected batteries on Mount Defiance, General St. Clair was forced to evacuate the position, and Ticonderoga again passed into the hands of the British. In 1777, after Burgoyne's defeat, it was dismantled, and though the British again occupied the position in 1780, it never became of value as a fortress. After the Revolution, the work were allowed to crumble and decay, and now the fortifications built at such fabulous expense are but a collection of picturesque ruins. The illustration conveys a better impression than can any description of their appearance. The outlines of the walls and ramparts can be traced, the walls of the officers' quarters are still comparatively sound, and a vaulted chamber, variously called the bakery or the powder-magazine, is still accessible. But when it is remembered that up to a few years ago, the cut stone and brick of the fortifications and barracks were carried away by the vessel-load, to build new villages on the shores of Lake Champlain, it is easy to see that only the most vivid imagination can reconstruct any considerable portion of "Old Fort Ti."

CHAPTER VI.

The Journey down Lake Champlain.

FOR beauty of scenery, Lake Champlain is little, if at all, inferior to Lake George. Everything is on a larger scale, and though there is nothing to compare with the lovely views through the Narrows or among the islands of the smaller lake, there is more of grandeur and expansiveness, and the combinations of mountain and water view afforded at various points on Lake Champlain are among the finest of American natural pictures.

This lake, discovered and named in 1609 by Samuel de Champlain, lies between the States of Vermont and New York, and has a length of 130 miles from Whitehall at the southern extremity, to its northern outlet. It varies in breadth from half a mile to ten miles, and in depth from 50 to 280 feet. Among the rivers that flow into it are the Chazy, Saranac, Ausable, and Boquet on the west; the Winooski and Missisquoi on the east. The lake discharges into the St. Lawrence River, through a river known by various names, as the Sorel, St. Johns, or more generally the Richelieu. The first forty miles of the passage northward from Whitehall is more like a ride upon a river than a lake, as in this portion it often narrows to less than half a mile in width, and in some places to fifty or sixty rods. The boat glides over the even surface of the lake; the woods, hillsides, and farmhouses are in full view; a fresh, balmy air floats from the pastures and hilltops to the waters of the lake; there is none of the monotony of a sea voyage, none of the pitching and tossing experienced on the great western lakes, but perfect comfort, easy motion, reviving air, constant changes of view, and most enchanting scenery. All these make a sail from Whitehall to Ticonderoga more like the motions of fairies wafted through realms of beauty, than the ordinary locomotion of mortal men. South Bay is on the west side of the lake, about one mile from Whitehall Landing; and near here, at a bend in the lake, known as the "Elbow," is "Put's Rock," where Major Putnam with a small body of men opened fire upon five hundred Indians who were in their canoes upon the lake, a few days before Putnam was taken prisoner at Fort Ann. Arriving opposite the outlet of Lake George, we shall see on our left and towering above us, Mount Defiance, still crowned with the remains of the works erected by Burgoyne in 1777, and beyond, across the mouth of the creek, the promontory and ruins of Ticonderoga. On the other shore of the lake, to our right, stands Mount Independence, where, in 1777, St. Clair had works connected by a pontoon bridge with the main position at Ticonderoga. In a moment we land at the little pier beneath Mount Defiance, and receive on board the passengers from Lake George. Proceeding north from Ticonderoga, the Lake gradually spreads out and becomes wider, and the scenery increases in grandeur. To the right we have constantly before us the green hills of Vermont, surmounted by the lofty bulk of Mount Mansfield; to the left, the rugged shores of New York, with the purple peaks of the Adirondacks in the distance, and before us the blue waters of the lake, studded with emerald isles. Twelve miles above Ticonderoga the lake narrows again, the shores of Addison closing in on the right, while Crown Point projects from the left, leaving only half a mile of clear water between. But on rounding Crown Point, the great Bulwagga Bay widens to the left, at the head of which is Port Henry, a landing place for the steamers, whence the Crown Point of history is visited.

Crown Point and Its History.

The importance of this point to the control of the lake was early recognized by the French, and in 1731 the first work, a pentagonal star-fort, with bastioned angles, was erected here and named Fort St. Frederic, in honor of Frederic Maurepas, the premier of France. The outer wall of limestone enclosed barracks, a church, and a bomb-proof tower. The French designed to establish here a province to be attached to the Canadian domain, with Point de la Couronne as its capital. In 1759, after Lord Amherst captured Ticonderoga, the French peaceably abandoned Fort St. Frederic, which had become untenable, and Amherst began here the construction of a first-class fortress, which eventually cost the British government ten millions of dollars. By a fire in 1773, the works were seriously damaged, and two years later, with their armament of 114 guns, they fell into the hands of Warner and his "Green Mountain Boys." The next year, the Americans, retreating from the disastrous attack on Quebec, wintered here, and in 1777 Burgoyne made this his depot of supplies. The peninsula is a mile wide, and is a solid mass of limestone, thinly covered with earth, and the remains of the works still extant, show that they must have been of wonderful extent and strength. The ramparts were half a mile in extent, 25 feet high and of the same thickness, with bastions, ditches, curtains and glacis outside, and enclosed a broad parade and massive stone barracks, whose walls still stand. A covered way led from the northeast angle to the edge of the lake, where a well ninety feet deep and eight in diameter is cut in the solid rock. The ruined ramparts are now covered with thorn apple bushes, of a kind peculiar to this spot alone, and said to have been brought from France, which in their season are aflame with crimson fruit. The ruins of the old French Fort, St. Frederic, stand on the steep precipice overlooking the lake, and 200 yards northeast of the newer British works. Near at hand are the remains of the French settlement, which history and tradition inform us, was at one time a village of 1500 inhabitants, with stores, paved streets, gardens and vineyards.

Opposite Crown Point is Chimney Point, where the French, in 1631, made their first settlement in this vicinity, just 100 years before they began to fortify Crown Point. They named it Point de la Chevelure. Fort Henry, at the mouth of Bulwagga Bay, north from Crown Point, is noted for its expensive iron works, the supplies of which come from the vast deposits of magnetic ore in the mountains to the northwest.

Down the Lake to Burlington.

From this point northward, we keep the green shores in view on either hand, with the mountain ranges for background, and soon turn into Northwest Bay, on the New York shore, where we land at the little town of Westport, whence stages run to Elizabethtown, Keene and the Saranac Lakes. Vergennes is soon visible on the east side of the lake, seven miles from the mouth of Otter Creek, which here empties into the lake. This town has special advantages for shipbuilding; and here the flotilla was built and equipped, which captured the British fleet at Plattsburg. Thirty miles north of Crown Point, on the west of the lake, is a geological curiosity known as Split Rock. Near the light-house a point runs out into the lake, at the end of which there is an island of half an acre or more in extent, separated from the main land by a fissure fifteen feet wide. The water flows through this fissure; and in it soundings have been made five hundred feet without finding bottom. Several theories have been broached to account for this formation, but none of them are perfectly conclusive. Just above Split Rock, is Essex, a pleasant village on the west shore. At this point the lake grows wider, giving greater room for navigation; and eight or ten miles above Split Rock the lake is five miles wide. At the town of Willsborough, eight miles north of Split Rock, is the mouth of Boquet River, a stream which rises in the Adirondack Mountains, and is the outlet of some of the most attractive ponds found in that range. Our course bears hence to the eastward, and soon we see Shelburne Bay, on the Vermont shore, the winter quarters and ship-yard of the Champlain steamboats. And just here is a favorable point to say something about these steamers. One would scarcely expect to find on this little inland sea, boats comparing favorably as to size, speed and luxury, with the floating palaces of the Hudson River, or Long Island Sound; yet such is the case. The boats Vermont, Adirondack and Champlain, with their handsome saloons, their sumptuous tables, their uniformed officers and crews, their broad expanse of open deck with armies of cane-seat chairs, their comfortable and nicely furnished state rooms, and their elegantly carpeted and richly furnished saloons, would do no discredit to either of the celebrated routes named above. As we stand across the widening lake to the eastern shore, we see Rock Dunder, a sharp cone, thirty feet high, rising abruptly out of the water. It is related that in the war of 1812, a British man-of-war fiercely cannonaded it, suspecting it to be a Yankee infernal machine. Near the middle of the lake are the Four Brothers islands, called by the French Isles des Quatres Vents, and soon we see the houses and spires of Burlington, at the head of the bay of the same name, on the eastern shore. A lighthouse on Juniper Island, and a breakwater that protects the shipping in the harbor, are the objects that are passed in approaching the landing, and the first thing that particularly attracts our attention is the immense quantity of lumber piled in huge masses along the wharves.

The City of Burlington.

Burlington is the largest and wealthiest city in Vermont, and is often called the Queen City. Its location, on a long sloping hill, whose foot is laved by Lake Champlain, gives it a magnificent outlook upon the beautiful water, stretching ten miles in width, and beyond the range of vision north and south. From the west windows of almost any building—so regular is the rise of the ground—a fine view is gained, but the best is perhaps from the dome of the University of Vermont. This institution, one mile from the shores of the bay, is 370 feet above its level. The three halls of the University have been united in one building, surmounted by a tin covered dome. The panorama presented to the eye from this point is truly wonderful, Lake Champlain, the Green Mountains and the Adirondacks being in sight, and over sixty mountain peaks distinctly visible. Beautiful drives stretch away in every direction; and the billowy mountain ridges, swelling into countless pointed waves, and scooped into deep hollows, abound on every side. But a short distance further inland, in the rural cemetery overlooking the Winooski, or Onion River, is the grave of Ethan Allen. The marble obelisk, supporting a heroic size statue of Allen, dedicated July 4, 1873, bears the inscription:

THE
CORPOREAL PART
OF
GENERAL ETHAN ALLEN
RESTS BENEATH THIS STONE.
THE 12TH DAY OF FEBRUARY, 1789,
AGED 50 YEARS.
HIS SPIRIT TRIED THE MERCIES OF HIS GOD,
IN WHOM ALONE HE BELIEVED AND STRONGLY TRUSTED.

BURLINGTON.

A short distance beyond, in a deep valley, are the falls of the Winooski, utilized for the propulsion of flouring mills, around which a lively village has sprung up. These falls are very romantic and striking, and in high water quite majestic in their proportions. Mount Mansfield, 4279 feet high, lies 20 miles to the northeast of Burlington; and Camel's Hump, 4183 fee', the same distance to the southeast. Conveyances may be obtained at Burlington for both these mountains. High Bridge, Howard's Summit, and Point Rock Institute are the places of interest which all travellers who can spare the time want to see. For this purpose many stop over night, and get a few hours in the morning to drive about the city and the suburbs. Excellent accommodations for guests are found at the American Hotel, managed by Mr. H. H. Howe, and at the Van Ness House. kept by Barber & Co., a new brick structure of considerable capacity and first class in its accommodations.

A Magnificent Lake View.

The view across the lake at sunset from either of these houses is worth a journey to Burlington in itself. The sun sinking below the misty peaks of the Adirondacks in the hazy distance, sheds a golden refulgence across the sparkling expanse of intervening water. The green islets on the bosom of the lake are illumined till they gleam like emeralds in the warm glow; and as the sunlight fades out and the darkness slowly settles down, the scene is bathed in an ever-changing radiance, turning from gold to crimson, from crimson to purple, from purple to violet, and from violet to the bluish-black of the early evening. And then the stars come out, one by one, till the firmament above is sprinkled as with silvery dust, and every particle of this shining dust is duplicated in the liquid mirror below. Then the moon rising over the great ridge to the east blazons a broad pathway of frosted silver across the lake, and turns each projecting rock, each lone dead tree into a silver milestone to mark the way. Such a scene witnessed once, can never be forgotten, and its memory is worth a summer of ordinary sight-seeing. The city is neatly built and regularly laid out, with several fine churches, an imposing city hall, fronting on a little green near the centre, a custom house and post office building, and an elegant court house but a block or two away. The lumber mills are the city's chief source of wealth. Some 50,000,000 feet are annually exported, and the business is constantly increasing.

Across the Lake to Plattsburg.

Leaving the wharf at Burlington, we steam nearly straight across the lake to Port Kent. South Hero — the largest island in the lake, and which with North Hero comprises the county of Grand Isle, Vermont — is seen right ahead, and passing west of this island, Valcour Island appears upon the left. Just south of Valcour Island was the first naval engagement of the Revolutionary war, on the 11th of October, 1776, between the American fleet, commanded by Benedict Arnold, and the British under Gov. Carleton and Capt. Pringle. Valcour Island is one of the largest and handsomest of the islands in Lake Champlain, excepting North and South Hero Islands, which are like continents almost in their extent, being laid out in farms of large extent, and having one or two considerable villages within their limits. Valcour is lofty, wooded to the water's edge, and surrounded on all sides by deep water. On the westerly side is a handsome stone lighthouse, recently built by the United States government, as an aid to the navigation of the lake. The experiment of founding a free-love community was tried in 1874 on Valcour Island, and quite a number of colonists went into the scheme, but as is usually the case with such experiments, internal dissensions and quarrels over the property involved, broke up the arrangement. Port Kent is a small village whence considerable iron ore is shipped and whence stages run to Keeseville, the Adirondacks and the Ausable Chasm. After leaving Port Kent, we pass

7*

through the narrow but deep channel between Valcour Island and the west shore of the lake. The wreck of the Royal Savage, sunk in the engagement mentioned above, still lies in the water of the Island, and is frequently visited by parties of the youth of the neighboring shore, who dive to the sunken hulk for "relics." North of Valcour is Crab Island, on which the Americans had a small battery in the battle of Plattsburg, and where the sailors killed in the fight were buried. Passing this island, we turn to the left and enter Cumberland Bay, a noble expanse of water, nearly land-locked by Cumberland Head, which projects far out into the lake and encloses the bay on the north and northeast. On its extreme point is a lighthouse, and at the head of the bay and mouth of the Saranac river which empties here, is the town of Plattsburg. A long mole or breakwater, with a small lighthouse on either end, protects the open side of the harbor, and within vessels can safely lie at any time.

The Lower End of the Lake.

Leaving Plattsburg and its neighboring attractions for future mention, we will continue down the lake. It seems odd to any one but a "native" to speak of going north as "down the lake," yet as Nature has willed that Champlain should discharge its waters into the St. Lawrence, north is "down" and we must submit to it. We round Cumberland Head on our left as we leave the harbor, and pass between the west shore and Grand Isle. Fifteen miles north, Isle La Motte, on which the French in 1665 built a fort, rises on our right, and to the east of this, the peninsula known as Alburgh Tongue makes down from the north on the eastern shore, enclosing a large and beautiful expanse or arm of the lake, known as Missisquoi Bay. The large island of North Hero lies directly in its entrance, leaving only a channel on either side. Ten miles further north, or twenty-five from Plattsburg and 130 from Whitehall, we reach Rouse's Point, the end of our steamboat voyage. Rouse's Point is a small and unattractive village, noted only as a railroad centre and as the frontier post, where Uncle Sam's custom officers inspect the baggage of tourists coming from Her Brittanic Majesty's dominions. Here the waters of the lake are discharged through the broad Richelieu or St. Johns river, seventy miles long, which empties into the St. Lawrence below Montreal. Who Rouse was, or why he located his point here, history does not inform us. It does tell us, however, that the fortification — in an unfinished and decidedly demoralized condition, a mile north of the village — is Fort Montgomery, and was built to command the Richelieu river, with 164 guns. After work had progressed on it for some time, it was found to stand on British soil, and only a generous change of boundary gave it to the United States. The Western Division of the Central Vermont Railroad, diverging westward from the main line at St. Albans, Vt., and running thence to Ogdensburgh, here crosses the river on a pile bridge a mile long, with a floating draw 300 feet in length. The water is very clear here, and from the car windows droves of fish can be seen inquisitively smelling of the bridge piles, and apparently waiting to be hooked from the windows of the house on the bridge, built for the accommodation of the draw tenders.

CHAPTER VII.

Plattsburgh and the Ausable Chasm.

AUSABLE is a word frequently heard in and about Plattsburgh The river of that name, taking its rise among the Adirondacks empties into the lake a few miles below the town, and the famous chasm — also called "Walled Banks of the Ausable" — is one of the most noted objects of interest in its vicinity. That it is worthy tenfold the fame it now enjoys, this narrative will soon make evident. But first, we must look about Plattsburgh a little, and make preparations for a trip to the chasm. The town proper is noted for two things, one in the past, the other in the present. The first is the battle which in 1814 did so much towards settling our last open disagreement with the British lion; the other is Fouquet's Hotel, which has done and is still doing so much for the comfort of visitors to this section. Taking the subjects in chronological order, we will read up a little on the history of the famous battle of Plattsburgh. The

British, in the war of 1812, looked upon Lake Champlain as one of the easiest as well as most direct routes for their invasion of the States from their Canadian possession. In the former assumption they were somewhat "out," as the sequel shows. The Americans had resolved to contest the supremacy of the lake, and both sides put forth their utmost energies during the Summer of 1814 in preparation. There was the greatest despatch in getting vessels ready for defence. The Saratoga, which carried twenty-six guns, being the largest American vessel on the lake, was built at Vergennes, and launched on the twentieth day after the first tree in her frame was brought from the forest. Capt. McDonough, the commander of the American fleet, anchored in Cumberland Bay on the 3d of September, and waited for the fleet of the enemy. His flotilla consisted of the Saratoga, the Eagle of twenty guns, the Ticonderoga of seventeen, the Preble of seven, and ten gunboats. On the morning of the 8th, the British fleet rounded Cumberland Head and advanced to the attack. Commodore Downie was in command, and his vessels were the Confiance of thirty-eight guns, the Linnet of sixteen, the Chub and Finch of eleven guns each, and twelve gunboats. The total strength of the Americans were 86 guns and 852 men, while the invaders numbered 95 guns and 1095 men. As the hostile fleet approached, McDonough knelt on the deck of the Saratoga, surrounded by his officers and men, and invoked the aid of the God of battles; then gave the signal to begin the action. The Saratoga and Eagle opened, when Downie's flagship, the Confiance, closing in upon the Saratoga, swept her decks with a tremendous broadside, and the Linnet by an advantage of position was enabled to rake her from stem to stern. But the brave McDonough kept up his fire till his whole starboard battery was disabled, when, by a skillful manœuvre, he turned about and opened such a terrific fire from his port battery that he compelled both his antagonists to strike their flags. Meantime, the Eagle had captured the Chub, the Ticonderoga had disabled the Finch, and after 2½ hours of steady cannonading, the whole British fleet had surrendered. Commodore Downie had been killed and McDonough had been disabled by two severe wounds. Sir George Prevost, meanwhile had attacked the land position of the Americans across the peninsula, between the Saranac and the lake, with 14,000 veteran British troops, and had vainly attempted to storm the rude earth-works by fording the river in three columns. The defeat of the naval forces, and fears of being surrounded by the rapidly gathering militia of New York and Vermont, led Prevost to precipitately retreat in the night, leaving his stores and wounded behind. His loss in this engagement was 2000; that of the Americans less than 150. General Macomb, who commanded the Americans, had but 1500 regulars, 700 New York and 2500 Vermont militia. The exultation of the Yankees over their victory was immense; McDonough and Macomb were loaded with honors and substantial rewards, while the chagrin of the British found vent in the degradation of Prevost, which Downie, being dead, did not partake. The scenes of the land and naval engagements are still pointed out to visitors to Plattsburgh, and the site of the American batteries are still visible.

So much for the memories of the past. Now for the realities of the present.

Fouquet's Hotel and its Attractions.

Fouquet's Hotel is the present feature, *par excellence*, of Plattsburgh and has given it more celebrity than any other one feature. No person visiting the Adirondacks — which in recent years have become no less a fashionable resort than a sanitarium and a paradise for the sportman — fails to stop at Fouquet's going or coming, to test the wonders of its *cuisine*, the luxury of its cool, fragrant bedrooms and sleep-wooing beds, and to revel in the balmy enchantments of its delightful flower-garden. From Fouquet's, by teams, which can be had on application, from the stables, sight-seers are conveyed to all points of interest in and about Plattsburg. This house

FOUQUET'S HOTEL.

has been known to the travelling public for more than seventy years. The family of Mr. Fouquet has met with great success in hotel keeping, having an appreciative sense of what travellers want, and providing accordingly. In June, 1864, the house standing on the site of the present beautiful structure was burned, and the present hotel was erected a year or two later. The late proprietor, Mr. Louis M. Fouquet—a son of Mr. D. L. Fouquet and grandson of Mr. John L. Fouquet, who in 1798 opened the first public house on this site—was a gentleman by instinct and practice; a man of intellect and culture, a poet, an artist, and a lover of all that is beautiful in nature; to which he added a never-failing thoughtfulness of the needs and wishes of his guests, and a personal interest in their comfort. He died May 26, 1875, and the house was purchased by Apollos A. Smith & Co., now kept by Messrs. Smith & Martin, well-known to Adironack visitors by their management of "Paul Smith's."

The present house accommodates 150 guests, its rooms are large, well ventilated, refitted and re-furnished, while the table is famed for its exquisite cookery and its perfect attendance. The house has always been the favorite resort of the United States officers stationed at the port, and MacDonough, Worth, Wool, Bonneville, Magruder, Hooker, Kearney, Richetts and "Stonewall" Jackson, have quartered here during their service at Plattsburgh.

The grounds and fragrant flower garden afford a most agreeable retreat. The broad piazzas on two sides of the house, and the promenade upon the roof, give a wide view of the lake, battle-ground, the scene of the naval engagement, the village, the surrounding country, and the mountains on every side. The house, by its beautiful and spacious grounds, furnishes safe and pleasant accommodations for ladies and children through the summer. There are large brick stables on the grounds, intended for the accommodation of guests who bring their own horses and carriages, as well as to furnish teams for rides and excursions.

FOUQUET'S GARDEN.

The Town of Plattsburgh.

The Plattsburgh of the present day is a thriving town whose business prosperity is due to its saw-mills on the Saranac river at the rapids near its month, and to its extensive lake commerce. It has about 6,000 inhabitants and is the capital of Clinton county. The court house and other county buildings stand on Margaret street, near the little park at the centre, and near by, on the same street, are Trinity Church and the Custom House. There are several handsome churches, and many of the streets leading from the bay over the high plateau back of the business centre, are lined with handsome trees and spacious grounds, while elegant residences look out from their environment of trees. The United States barracks, a frontier post of considerable importance, and during the war a receiving depot for troops, are on a sandy bluff a mile south of the village, and the bugle call which announces "reveille," "retreat," and "tattoo," is one of the familiar sounds of Plattsburgh. There are several very pleasant and interesting drives in the vicinity of Plattsburgh. That around Cumberland Head affords splendid views of the lake and the scene of the battle of 1814. Another is to the town of Dannemora, sixteen miles northwest of Plattsburgh, in which is situated the Clinton Prison, as it is called. Here is an elevation 1,700 feet high; but the ascent is so gradual that in riding up from Plattsburgh it is scarcely observed to rise at all. This is visited chiefly for the beautiful view that is afforded of the surrounding country, — the Green Mountains in the east, Lake Champlain stretching to the southeast, the winding course of the Saranac to the south, and the high Adirondack peaks to the westward. Five miles beyond, in the town of Dannemora, is Chazy Lake, a little gem, set in the most picturesque and beautiful surrounding. This lake is about four miles long by two wide, and is a favorite resort of the sportsman as well as the admirer of natural scenery. This lake is nowhere surpassed as a place for fishing. Trout weighing twenty pounds have been taken from it. The road to the lake was built with great care, at the expense of the State, and is in good condition. Near Chazy Lake stands Lion Mountain, sometimes called on the maps Lynn Mountain; Bradley Pond is two miles west of Chazy Lake, and west of this there is a path running three miles farther to the Upper Chateaugay Lake. West of the Upper Chateaugay Lake is Ragged Lake, six miles long and half a mile wide. From the town to the mouth of the Ausable, and back along the lake shore, is another favorite drive and one of the finest on a clear morning that can anywhere be enjoyed. As this is a part of the route to the Ausable Chasm, and as we have been all this time getting ready for the trip hither, we will suppose ourselves stepping from the verandah of Fouquet's into one of Ransom's easy carriages, draw by a pair of spanking bays, in the clear, cool early morning, bound for the Chasm.

The Drive to Ausable Chasm.

Our course leads us out through the southern part of the town, across the sandy plain by the barracks, past the Catholic cemetery with its odd monuments, in sight of the Saranac river, with its amber waters churned into foam as it pours over the rocky ledges, and millions of logs, floated down from the Adirondack woods and ponds, stranded and waiting for the next freshet to send them down to the booms above the Plattsburgh saw mills. A little further on our road descends to the shore of the lake, which it follows for several miles. The panorama is both grand and beautiful. To the left is the clear, blue lake, stretching away almost beyond the line of vision, with the Vermont hills forming a low purple wall in the background. Near at hand, the liquid expanse is broken by Valcour Island, with its dense green foliage, and at our feet lie occasional fishing boats, about which are fishermen busy with their nets. Along the lake shore, at the outer edge of the road, a line of poplars, tall and erect, set before the war of 1812, stand like sentinels watching the border.

To our right are smooth, rolling farm lands, with solid stone houses and mammoth barns, telling of fertile acres and abundant crops. In the distance the Adirondack peaks pierce the clouds, and their grim wall, softened by the distance to a most unsubstantial looking blue, shuts in the scene Before us Mount Trembleau stands like a grim giant to bar our way, and across the lake a great mass like a lion couchant indicates and justifies the name of Lion Mountain which it bears. At intervals of a few miles, old landings project into the lake, where, before the days of the railroad, sailing vessels and steamers landed, and whence plank roads led back into the country, but which are now obsolete, and whose docks and storehouses are falling to pieces. We ford the Ausable river near its mouth, and opposite a lovely butternut grove, a favo ite picnic resort in summer. The river is now but two or three feet deep at this point, but in the Spring it tears and roars along all over the surrounding bottom lands, as shown by the floodwood and gravel deposited here and there on the meadows We follow the general course of this stream two or three miles up, till on rising a hill we see a turn in the road to the right, a plash of falling water, a stone bridge, and on the other side a gray stone house on which is painted "Chasm House." Here our driver pulls up, and informs us that here is the head of the chasm, and that it extends a mile or two close beside the road over which we have come. All we have seen has been a dense thicket, mostly cedar—not a sign of a glen or a stream. And no wonder. We might have walked through this thicket, and until we reached its very brink, or perhaps stepped off into the narrow, vertical crevice, we should not have seen the chasm.

The Chasm and its Wonders.

We drive up to a small wooden building called the Lodge, which we enter, leaving our wrappings and "traps" in the carriage to be regained at the lower end of the chasm, to which our driver proceeds, while we make the underground

THE route through the Chasm described in the adjoining pages, begins at the "Upper Entrance" near the bottom of the Map, and continues upward to "The Pool" near the upper right hand corner.

MAP

OF THE

Ausable Chasm.

[Taken by permission from Stoddard's "Adirondacks."]

trip on foot. We pay half a dollar a piece for a ticket of admission, are shown to the head of a flight of stairs on the chasm side of the house, and start on our journey. There are 100 steps in this almost perpendicular staircase, and we descend with a good deal of trepidation, and stand on the solid rock at the bottom with a sense of relief. Here we find ourselves looking at two prodigious perpendicular walls, 75 feet high, one on either hand. Almost at our feet, the shelf of smooth rock on which we stand gives place to vacancy. Stepping to the edge we look and see the river roaring over its rocky bed a few feet below us. We proceed up stream a few rods, and turning a corner of the rocky wall we are face to face with and at the foot of a cataract very like the American fall at Niagara, in all save grandeur. It is the Birmingham Fall, sixty feet high divided in the centre by a tower of solid rock on which rests the pier of the bridge which we have seen from above, and which spans the river just at the verge of the fall. We gaze at the foamy cataract for a few moments and then begin our journey down stream, through the wonderful glen. For the first few minutes we proceed on the right bank of the stream, along the same rocky shelf or ledge to which we first descended (past the foot of the stairs, and the Horseshoe Falls, where the river tumbles over a semi-circular ledge some ten feet in height, and then makes an abrupt turn to the right) and then cross the stream on a wooden foot bridge resting upon the ledges on either side and a rocky islet in mid-stream, whence we gain a splendid view of the Birmingham and Horse-shoe Falls.

From this point, we climb by a most trying flight of stairs, to the summit of a lofty projection of the left bank of the river, which bars further progress along the ledge, then crossing the point on the top of the earth, we descend on the other side, by another flight of steps. This projecting tower bears the name of Jacob's Ladder. A minute description of the rest of the trip through this wonderful glen, would be merely a chronicle of steps ascended and steps descended, some 700 in all; of long galleries of plank anchored to the rocky wall traversed, by which we continue our journey where the ledge of rock ceases to give a foothold; of one long stretch where the waters in the Spring freshets have cut a passageway in the stony sides of the chasm, barely wide and high enough to permit of our crawling like insects, while the honeycombed depressions in the rock, filled with the cards and address envelopes of previous visitors, give the spot the fitting name of "The Post Office," and last and most delightful feature of all, a glide down stream in a little boat, through the famous Flume.

To generalize the tour, it is only necessary to say that the chasm is a rift in the massive rock—of Potsdam sand-stone, —from ten to fifty feet wide, and from sixty to 200 feet deep. Through the bottom, the Ausable makes its way, now pouring noisily over a little fall, now gliding rapidly over a ledge, now expanding into a placid pool, and now surging through a narrow gateway, where the eternal walls seems disposed to close, and forbid entirely the passage of the stream. The Flume, already mentioned, is a long, smooth stretch of water, between walls almost as smooth and straight as if carved and polished by the stone-cutter's art. Here the gorge is very narrow; so narrow that you can almost touch the walls on either hand, and the water so deep as to appear perfectly black. Through this gloomy pass, recalling somehow, reveries of Venetian canals, the Bridge of Sighs and the passage of black gondolas with prisoners for the gloomy subaqueous dungeons of the Doges, we smoothly drift in the small batteau to the lower end, where the walls expand; the grateful sunlight, so long excluded, pours down upon the water, the banks recede and slope gently to the river's brink in a little sandy beach, and the broadening stream roars and plashes over a little rapid just below. We land on the beach, and a flight of stairs ascends to the top of the bank, where our carriage waits. At the head of the stairs is a small house, where mild refreshments, as ginger pop and soda, are dispensed, and where excellent photographs and stereoscopic views of the marvellous Chasm are for sale, at prices fully as low as are charged for similar articles in Boston or New York. A few of the chief objects of interest on our way down the Chasm, are Jacob's Ladder, a bold, projecting tower, around which the river makes a sort of gooseneck curve; the Leaning Tower, the Tower of Babel, the Pyramid, Point of Rocks, Point Surprise, the Sentinel and the Broken Needle, all lofty columnar forms; Mystic Gorge and half a hundred of other nameless rifts or lateral fissures, branching off at right angles to the main cleft, into its solid walls; Cathedral Rocks, between which a flight of 210 steps rises to the bank above;

Cathedral Rocks.

Table Rock, a broad flat plateau of stone at the foot of Cathedral Rocks, and no end of attractions dignified by the name of his Satanic Majesty — the Devil's Oven, the Devil's Slide, the Devil's Chimney, the Devil's Punchbowl, the Devil's Pulpit, and Hell Gate. The Devil's Oven is merely a large cave in the side of the Chasm, a little below Jacob's Ladder and on the opposite side. It has been formed evidently by the gradual disintegration of the layers of soft rock, which have crumbled into little cubical blocks and fallen down, so that there is quite a sloping pyramid of them leading up to the entrance of the oven. His Slide is one of the transverse fissures, which extends in a regular slope from the bottom of the glen up to daylight above; his Pulpit and his Chimney are projections of rock from the top of the cliffs, resembling the objects named; and his Punchbowl is a deep pool formed by a turn in the great cleft through which the river runs, and instead of punch it holds the clearest and purest of water. Why all these notable objects should be surrendered to the custody of the Prince of Darkness it is hard to say, unless it be to deter the visitor from any contemplated wickedness by the fear of drowning in the Punchbowl or roasting in the Oven. But travel where you will, the most wonderful and grand objects in nature are named, with wonderful unanimity, after the Devil. Until within a few years, the Ausable Chasm, though a local wonder, was little known to the outside world, and few visits were paid to it. About 1868, the stairs at the Cathedral Rocks were built, and visitors

CATHEDRAL ROCKS.

who had hitherto scrambled down as best they could near the Devil's Oven, were provided with a safe and comfortable means of entering the Chasm. In 1873, a party of Philadelphia capitalists purchased the land on the right bank of the glen, built the lodge, the stairways, galleries and bridges, and put in the boat. They also built in the summer of 1874 a new hotel on the high ground overlooking the lake, the river above the Falls and the surrounding country. It is called the Lake View House, and is a handsome three-story structure, with a tall tower, large, airy, well furnished rooms lighted by gas, and all the requisites of a first-class summer hotel. With the present facilities, the Ausable Chasm is one of the most accessible and easily visited places, and certainly none will better repay the tourist for a day's time and a few dollars expense.

Fo
Plac

This fold-out is being c

CHAPTER VIII.

The Adirondack Region.

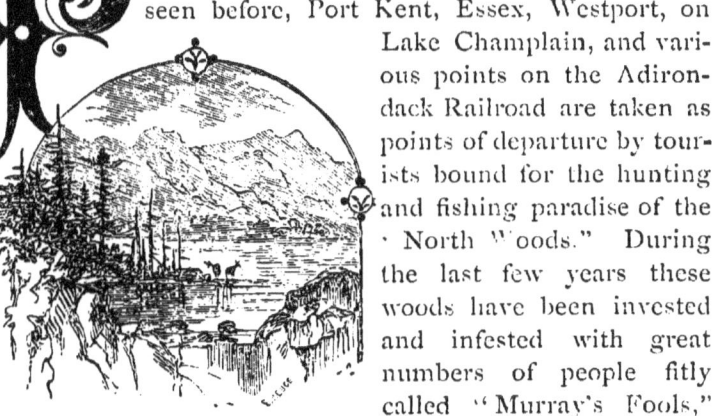

LATTSBURGH is the most convenient point from which to penetrate the Adirondack region from this direction though as we have seen before, Port Kent, Essex, Westport, on Lake Champlain, and various points on the Adirondack Railroad are taken as points of departure by tourists bound for the hunting and fishing paradise of the "North Woods." During the last few years these woods have been invested and infested with great numbers of people fitly called "Murray's Fools," who have rushed hither imagining that it was the thing to do, without the slightest intelligent idea as to the nature of the country or what is required for the proper enjoyment of the resort. They have come hither attired in watering place style, the men with most elegant guns, rods and flies, expecting to shoot deer and catch trout from the piazza of a hotel, and to wear their silk hats, white trousers, primrose kids and patent-leather boots while doing it; the ladies with beruffled and befrilled silk and muslin costumes and Saratoga trunks, bent on making a sensation—which they generally do. These silly people have brought the Adirondacks into disrepute by their piteous lamentations on their return to civilization of the discomforts they have endured and the disappointments they have suffered; but they have mostly given up trying to be woodsmen and wood nymphs, and have abandoned the Adirondacks to those who can appreciate and enjoy them; the experienced and sensible lovers of Nature in her wildest moods, and of the free life of the forest. The range of mountains known as the Adirondacks extends from the north-east corner of New York State, in a south south-westerly direction, occupying portions of Clinton, Essex, Franklin, and Hamilton Counties. It finds its greatest elevations in the western part of Essex County, which contains the highest peaks of the Northern Appalachian Chain, Mount Washington alone excepted. The sources of some of the streams which flow in different directions are often connected with each other, many of the lakes and ponds lying on the same plane. Most of these bodies of water vary in height above the sea-level from 1500 feet to 1731 feet, the latter being the elevation of Raquette Lake. The mountains are well covered with trees, — birch, beech, maple, ash, hemlock, spruce, fir, cedar, and white pine, in the higher lands, and along the courses of the streams almost impenetrable thickets of tamarack, hemlock, and cedar. The pine affords the most valuable timber, which is run down the various streams in the time of the spring freshets. Masses of magnetic iron ore of enormous extent have been found, which have led to the establishing of smelting works. The tour usually made by the casual visitor embraces the St. Regis and Saranac Lakes, with Paul Smith's and Martin's as the central points. The St. Regis lakes are the northermost of the chain which lies to the west of the mountain range. Taking an early breakfast at Fouquet's we take the train on the Whitehall and Plattsburgh Railroad, from the depot near the hotel, and proceed southwest twenty miles to Ausable Station, on the river of the same name, whence Concord coaches convey us three miles over a plank road to Ausable Forks, an iron mining and smelting village, with two small taverns, a telegraph office and a few stores. Here the north branch of the Ausable River, fed by Lake Placid and other ponds, joins the south branch flowing up from the Ausable Ponds, and the shallow, brawling stream, thence pursues a northeastern course till it passes through the marvellous Chasm, and empties into Lake Champlain. From Ausable Forks to Franklin Falls is 17 miles of hard mountain staging, and at the last named place, passengers stop for dinner. The plank road ceases here, and the rest of the journey is made in light mountain wagons, over roads which will compare favorably with country roads elsewhere. Our course is nearly due west, and we pass through the neat little post village of Bloomingdale, eight miles from Franklin Falls, whence we may take our choice of roads—to Smith's 10 miles, to Martin's 10 miles, and to Cox's 15 miles. We will take the first-named route, and after traversing what seems to us fully thirty miles of road, much of it recently cut through the virgin forest, but in very good condition, and accompanied all the way by a telegraph line, we shall be deposited on the long piazza at " Paul Smith's " tired enough to be contented with the most uncomfortable quarters, and hungry enough to devour without question the least inviting of fare.

Paul Smith's and Its Luxuries.

But our heroism will not be put to any very severe test. Those who have paid previous visits here know, and all who come here for the first time are lost in wonder at discovering that rooms as airy and large, beds as comfortable, table linen as snowy, silver as bright and viands incomparably better than those of the fashionable watering-place hotels are here to be found. A supper of lake trout, venison steak, waffles and honey, with excellent tea or coffee, in Paul Smith's cool, airy dining room, and a night's sleep in one of his large, clean chambers, are calculated to make the wayfaring man feel at peace with himself and with all mankind. Apollos A. Smith, whom everybody calls "Pol" or "Paul" for short, is one of the oldest and best known of Adirondack landlords. He settled here in 1861 on the shores of the Lower St. Riges, and soon his quiet cottage became locally famous as a resting-place and headquarters for sportsmen who penetrated the North Woods. The region grew famous, and Paul, having that born instinct which alone can make a successful hotel-keeper, enlarged his borders from time to time till now he has a fine three-story house with accommodations for 100 guests, large stables for his own horses and those of guests who come in their own turnouts, and a big boat house on the sandy lake beach, a few rods from the house. In this last building a hundred of the light, graceful and staunch boats in universal use on these waters are "pigeon-holed," with sterns to the water, reminding one of the scene in "Lucrezia Borgia," where that estimable lady displays her facilities in the amateur undertaking business. These boats are all identical in size and build, are numbered from 1 up, and resemble very much the famous Whitehall boats, except that they are lighter. From the boat house to the water's edge a sloping platform of plank enables the rower to slide his boat out of water. About the boat house and the stables, will be seen a number of men in rough serviceable garb, many of them young, and all straight as Indians and almost as brown; with athletic frames, brawny hands and open, frank countenances; these are the guides. They may be engaged for the service of parties desiring to camp in the woods or on the ponds in the vicinity, for fishing and hunting purposes, or to convey parties down the lakes to Martin's, Bartlett's, Cox's or elsewhere. A finer set of fellows one need not seek, if he but respect their sturdy independence and follow out the Golden Rule in his treatment of them. By far the pleasantest way of reaching the points enumerated above, is by boat through the lakes, though the stages make daily trips. To gain the best idea of this wonderful region we will take to the boats.

A Boat Trip Through the Lake.

We will take with us only a light travelling bag each, with a change of clothing and toilet articles. We shall have provided ourselves with some old clothes, the ladies with flannel or waterproof travelling suits; the men with tweed or other stout clothing, flannel shirts, and all with thick boots or shoes. Thus equipped, with the addition of the necessary guns and fishing tackle, we are ready to set out. The guides will provide boats and oars, yokes to carry the boats over the portages, and the necessary strength and experience for managing them. Setting out from Paul Smith's, we pull across the lower St. Regis, southwardly, to its outlet, a broad, deep creek, through which we glide, amid the fragrance-laden waterlilies, which almost impede our progress, so plentiful are they; past the pine-crowned points and the tamarack-skirted lowlands, out into Spitfire pond, whose waves are gaily dancing in the morning light, to the rustling music of a fresh breeze. Crossing this small but sprightly body of water, our boats glide through a narrow creek to the St. Regis Lake, a beautiful sheet, which gives its name to this entire northern chain. When we touch its southern shore, our guides leap out and drag the boats ashore. There is a "carry" or portage of a mile and a half, by a crooked trail, through a dense wood. Formerly the guides carried the boats by means of the yokes, which resemble those used by maple sugar makers to carry the buckets of sap, but now the boats are dragged by horse-power, on a kind of big sled or jumper, by which means "old Sanguemaire," a French half-breed, living on the southern side of the carry, earns an honest livelihood, We trudge through the woods, perhaps bagging a squirrel or two by the way, and again launch our skiffs on the waters of Big Clear Lake, and pull southward across its dancing waters. Soon we drift into its outlet, a broad quiet creek, now deep and sluggish, by reason of a mill dam below, which soon compels our guides to shoulder and carry the boats around it. As they march off in single file, with the boats on their backs, bottom up, they look oddly like huge mud turtles. Launching again, below the mill, we soon emerge from the winding stream, into the upper Saranac Lake, the largest and finest of the chain. We pull across its clear and lively waters to the western shore, where near the head, stands Cox's Upper Saranac Lake House, a fine new structure, with rooms for 100 guests, and a table supplied with the best of forest cheer. Here we shall find it convenient and expedient to dine, before attacking the eight mile pull down the Upper Saranac Lake. We start afresh after dinner, and traverse the length of the beautiful lake; past shores of bold, precipitous rock, fringed with evergreens; past sailing loons, whose slim necks alone emerge from the waves, and at which we fire and don't hit, ; past bays and creeks, and all the beauties of lake and mountain scenery, and emerge into the Saranac river, the outlet of this lake, and soon reach Bartlett's Sportman's Home, a long, rambling house, chiefly piazza, where a smoking supper and a night's rest awaits us. The river falls some 60 feet here, so in the morning our boats are carried around and we again start down the Saranac River, and float out upon Round Lake, a lovely sheet of water, smooth and clear as glass in the early morning, but apt to be fretted by winds later in the day. Our course changes here, and we pull across this lake, some four miles, in a northerly direction, and again follow the course of the Saranac river, 3 miles to the Lower lake of the same name. On the way we shoot a rapid of eight feet fall, if we are brave enough to stay in our boats and try it; if not, we walk around and see the guides do it. If we choose to stop and "drop a line" here, we shall be quite likely to catch some fine trout in these rapids. As we enter the Lower Saranac Lake, Ampersand Mountain looms on our right, south of the lake, and Saranac mountain on our left, West; straight ahead, but afar off, old Whiteface towers aloft, and in the distance, to our right, we get occasional glimpses of Mounts Seward and Marcy. Ampersand mountain may be ascended, if we care for a three hours' climb, and the view from its summit (embracing the lakes we have traversed, as well as Long Lake, to the South, and the Tupper Lakes to the West, the mountain peaks far and near, the Saranac valley, and the beautiful and sequestered Ampersand Pond, to the south of the mountain) is worthy the exertion. We are now passing a frowning wall of rock, rising 150 feet perpendicularly out of the water, and wooded at the top, and soon we see a guideboard bearing the words "Jacob's Well." Here is a splendid living spring of pure sweet water, very refreshing to our throats, after our lively rowing. Hence we lay our course down the lake, six miles in length, to Martin's. This lake is diversified, and most beautifully, too, by over 50 islands, of which Eagle Island is the largest. We skim lightly by the rocky headlands, and the tree-studded islets, and in due time we draw alongside the little wharf in a beautiful bay at the northeastern extremity of the lake, and directly in front of Martin's Saranac Lake House. Here we meet cordial welcome from William F. Martin, one of the very best fellows in this wilderness, and who, by his enterprise and courtesy has reared, from the beginning made in 1849, a house of three stories, accommodating 200 guests, and just as comfortable and well kept as any other hotel in this vale of tears. Martin is a thorough woodsman and a genial host, and it is a treat to sit on his piazza of an evening, looking out across the lake, and listening to his stories of the early days of hotel keeping in the Adirondacks,

Martin's, as has been stated, is about the same distance from Plattsburg as Paul Smith's — 37 miles — and is frequently taken as the point of departure for the tour of the lakes, instead of Smith's; in which case the route we have passed over will be reversed. The tourists who propose to visit Tupper, Long and Raquette Lakes, Mounts Seward and Marcy, and the Indian Pass, generally make Martin's their rendezvous, and start thence with their guides and supplies up the Saranac Lake. Martin's being on the regular stage road and having mail and telegraphic facilities, the same as those at Smith's, is in every way a desirable headquarters.

8*

A Visit to the Southern Lakes.

A reference to the map will show in the southwestern portion of the Adirondack region, a cluster of large lakes connected by the Raquette river. These are the Tupper Lakes, Big and Little, Long Lake, Forked Lake, Raquette Lake and a little chain of small ponds leading from the last named and terminating in Blue Mountain Lake. The tour of all these lakes is made from Martin's by two routes, one leading west to the Tupper Lakes, and the other south to the Raquette, Long and Forked Lake chain. Taking the western route first, we proceed up the lake and across Round Lake to Bartlett's; thence two miles west to the old " Sweeney Cary " now operated by the Daniels Brothers, who haul our boats across a neck of land three miles wide, and launch them in the Raquette river. Eleven miles of tortuous gliding bring us to the outlet of Big Tupper Lake, where Martin Moody, an old guide, has built a house with accommodations for fifty guests, and has a large congregation of sportsmen each summer.

Big Tupper Lake.

This is a beautiful sheet of water, surrounded by ridges of mountains and by primeval forests; stretching away seven miles to the south and spreading out from one to three miles wide, dotted with islands, bordered by beautiful bays with green valleys rising from their heads, and with all the placid serenity that marks the quiet of this great wilderness. On Bluff Island, in the northeastern part of the lake, is a precipice overhanging the water, known as the Devil's Pulpit; and at the head of the lake, Bog river makes its way to the lower level by a fall resembling a sheet of silver, over a rocky ledge. From Big Tupper, a seldom traversed route leads south, through Round Pond, to Little Tupper Lake, which is six miles long, with high, rocky shores, numerous islands and beautiful scenery. The Wolf Pond route, so called, passss from the outlet of Big Tupper, by a sweeping curve to the northeast, to Cox's on Upper Saranac Lake, thirty miles distant, by way of a chain of small ponds. From the same point a twenty mile trip, including eight miles of carries, brings us to Cranberry Lake, one of the largest of the whole region, being fifteen miles long and discharging through the Oswegatchie river, and another trip of fifteen miles brings to the most desolate and lonesome (but most prolific in game) of these waters, Mud Lake. It is seven miles south of Cranberry Lake, and is but four miles in circumference. Still another route is through Little Tupper and a chain of ponds to Long Lake. The usual route to Long, Forked and Raquette lakes, however, lies south from Martin's via Saranac and Round lakes to Bartlett's, thence by the Indian Carry of one mile to the Stony Creek ponds, three in number, connected by narrow passages, and thence through Stony Creek, a narrow, tortuous stream, for three miles to its outlet into the Raquette river, twenty miles above Tupper Lake. A striking panorama of woodland scenery is presented as we ascend the river to Raquette's Falls, six miles, where is a carry of a mile and a quarter, and where stands "Mother Johnson's," renowned for pancakes, and noted in every tourist's description. The old lady herself, after entertaining thousands of guests during her stay in the quaint old log cabin, and gaining a fame excelled by no one in the region, died in the early spring of 1875, and her body was tenderly carried by the guides some thirty miles down the Raquette and laid to rest.

Long Lake and Its Scenery.

Six miles more pulling bring us to the foot or northern extremity of Long Lake, which is simply a widening of the Raquette river, or else the river is simply the thread on which Long Lake is strung. It is a handsome body of water, fourteen miles long and a mile wide at its widest point. It has several islands, one of which supports a small inn, owned by John Davies, who also keeps the famous "Aunt Polly" tavern at Newcomb, thirteen miles southeast. He maintains communication between the two houses, by a romantic forest route, through five small ponds and Catlin Lake, a body of water three miles long, and famous for its camping grounds. Long Lake is surrounded by mountains, Owl's Head, Mount Kempshall, Buck Mountain and Blueberry Mountain, and on the east bank, three and a half miles from the head or southern end, is Long Lake Village, or "Gougeville," where is Kellogg's famous hostelry, and where several families of noted guides reside. There is a little church, a school-house, a store and post-office here, and many visitors to the Raquette region make their headquarters at Kellogg's. Round Island lies about midway of the lake, and is a beautiful little gem of living green. Little Tupper Lake is reached from Kellogg's by a tedious route of 10 or 12 miles, which passes through Slim and Mud Ponds, and consumes a whole day. Owl's Head Mountain is ascended by the aid of guides, and a fine view is obtained from its summit. Near Kellogg's is a floating bridge across the lake, and beneath a portion of it is passage for boats. Four miles from the village, the head of the lake is reached, and here the Raquette river enters it, over rapids necessitating a half-mile carry. Then it is fair paddling again for a mile up stream to Buttermilk Falls, which are generally considered the original of Murray's Phantom Falls. The descent is about 20 feet, over a rugged ledge, with boulders which churn the water into a froth. A short carry here, then a mile and a half of boating, another carry of a mile and a half, and we reach Forked Lake. This expansion of the Raquette river is an irregular and very romantic pond, three miles long, with wooded shores and some fine scenery. It boasts an attachment, Little Forked Lake, through which, and a string of ponds with intervening carries, a twelve-mile route to Little Tupper lies. This whole country is so filled with lakes and ponds that you can go anywhere you choose with a boat, alternately carrying the boat and having the boat carry you.

Raquette Lake and its Tributaries.

From this body of water a half-mile portage brings us to Raquette Lake, twelve miles long and from one to five miles wide; one of the most beautiful of the Adirondack waters, and destined to be the favorite, when hotels and other products of civilization shall shed their ameliorating illumination upon its dark waters, so to speak. Meantime, only enthusiastic sportsmen and hardy tourists find their way hither. Cary's Hotel, a rude and small house, is the only hotel on the lake, and nearly every one who comes hither camps out. Beach's Island, a beautifully wooded and symmetrical islet; Murray's Island, the camping ground of the celebrated preacher, and the numerous points which project from the shores are used for camping. The old State road from Crown Point to Carthage, which was formerly a travelled way through this region, passes Cary's, and it is only 14 miles to Long Lake Village, but of late years it is seldom used, and has degenerated into a mere trail. The lake is surrounded by picturesque mountains, and is deeply embayed on nearly every side; Marryatt's Bay, Eagle Bay, North Bay and South Bay being the principal arms of the lake. Little Tupper Lake is accessible from Raquette also, by an 18 mile route embracing Beach's and Salmon Lakes and two or three carries. Shallow Lake, reached by an inlet from the west shore, and a group of neighboring ponds are famous retreats for trout.

The John Brown Tract.

A pull of four miles up the Brown Tract Inlet, from the south-west point of Raquette Lake, and a carry of a mile and a half bring us to the upper or eighth of the Fulton chain of lakes, which extend southwesterly into the "John Brown Tract," so-called. The ardent hunter, who has not yet gained enough experience of the lakes and mountains, follows this route, which brings him into the Eighth Lake first, passes by portage to the Seventh, and can go directly by boating into the Sixth. The Sixth and Fifth are quite small; and there is a portage between Sixth and Fifth, and also one between Fifth and Fourth. Fourth Lake is the largest of the chain, and has a number of islands in it. The shores are high, and rise in rapid ascents. Hemlock grows down to the edge of the water; and in the undisturbed repose of the waters the fringes of foliage are clearly reflected. In the centre of the lake is a beautiful group of rocks known as Elba. There is a passage for boats into Third Lake, close by which Bald Mountain frowns down; and the passage continues open to Second Lake. Second is hardly distinguishable from First, there being a mere sand-bar separating them. The Adirondack Railroad will pass just north of these lakes.

The southwestern part of the Adirondack region, known as the John Brown Tract, reaches into Lewis and Hamilton counties, but is mostly included in Herkimer county. In area it is about twenty miles square, and is supposed by many to take its name from the hero of North Elba, but such is not the fact. John Brown was a merchant of Providence, R. I., and coming hither in 1792, bought this tract, which he divided into eight townships named Industry, Enterprise, Perseverance, Unanimity, Frugality, Sobriety, Economy and Regularity. In 1812, Brown's nephew, Charles Herreshoff, tried to found colonies in these model townships, and incurred great expenses for clearings, mills, etc. A large number of people from the seaboard attempted a permanent settlement, but many discouragements appeared to thwart them; work upon the tract was abandoned; Herreshoff suicided after a seven years' struggle with fate, and the solitude of the unbroken wilderness again resumed its sway. This section is only visited by hunters and fishermen, and they succeed in their designs upon the game and the fish.

The Eckford Chain of Lakes.

But after this long digression, we must get back to Raquette

Lake, whence, by the Marion river, the principal feeder of the lake, we make our way through Utowanna and Eagle Lakes, to Blue Mountain Lake, the gem of the southern Adirondack waters. Austin's Hotel, between Eagle and Blue Mountain lakes, is a favorite resort of sportsmen, and "Ned Buntline" (E. Z. C. Judson), has a lodge on the shore of the former. These pretty lakelets, the Eckford chain, as they are sometimes called, are the headwaters of the Raquette, and 'Mount Emmons, or Blue Mountain, which overlooks the lake bearing its name, is ascended by a trail on its western slope. It is 3,595 feet high, and its summit having been cleared of trees during the State survey of 1873, a magnificent panorama is spread out before the visitor who stands on its top. Blue Mountain Lake, by the route we have come, is fifty-five miles from Kellogg's, but as we have described a deep loop in onr journey, only five miles separates us from Kellogg's now. The mountain divide and South Pond, with a mile's portage are the obstacles. The mountain trail is arduous, but the guides often attempt it in preference to going back.

The Southern Adirondacks.

From Long Lake Village a weekly mail stage or private conveyance may be taken for a trip to the deserted Adirondack iron works, Mount Marcy and the Indian Pass. The State military road, previously mentioned, runs due east almost for forty-two miles, along a route alternating mountain, forest and lake scenery, and is sufficiently rough to give any one a wolfish appetite every five miles, to Root's, a famous resort for sportmen, with accommodations for forty or fifty guests. Thence the roads diverge; one to Ticonderoga, twenty-three miles southeast, passing Paradox Lake and Long Pond; another to Crown Point, eighteen miles northeast; a third south to Schroon Lake, nine miles, and the fourth north to Elizabethtown, twenty-two miles. But we stop nineteen miles short of Root's, at Tahawus or Lower Works. Here the Hudson or North river, a narrow creek at this point, was utilized to furnish power for the iron works, and the dam which once spanned it flooded the valley up to Lake Sanford, five miles above, and barges plied between the Lower and the Upper Works. The road north to the Upper Works or Adirondack, is picturesque and full of interest. The distance is eleven miles, five of which are along the shores of the mountain-walled Lake Sanford. Reaching Adirondack, a scene rarely to be witnessed in America is presented — a ruined village — and it was formerly a manufacturing village, too. The tall chimneys, the furnaces, the old school-house, the church and the dwellings of the workmen, all are abandoned. Only Moore's, a small and modest house of resort, is occupied. The history of this place is brief. In 1826, an Indian discovered immense deposits of iron, and a dam of almost pure ore across the river at this place, and reported his find to Messrs. Henderson, McMartin and McIntire who had iron works at North Elba. They secured the whole territory, built forges, furnaces and a road to Lake Champlain and began operations. A busy and thriving village sprung into existence, but the venture proved unsuccessful, the cost of getting the iron to market being too heavy, and Mr. Henderson, the head of the firm, being accidently killed in 1845, the Upper Works were in 1848 abandoned, and the Lower Works soon after, since which time the villages have gone to decay. Only the names of the three unfortunate speculators are perpetuated in Mounts Henderson, McMartin and McIntire, and Lake Henderson. The Preston Ponds lie two miles northwest, and six miles further in the same direction is the grand peak of Mount Seward, 4,348 feet high. Mount Marcy (called by the Indians Tahawus, the Cloud Piercer or Sky Splitter), the monarch of this region, towers to the height of 5,333 feet, to the northeast, and is ascended by a trail which necessitates twelve miles of arduous climbing. The ascent can only be made by strong and indomitable climbers, by the aid of guides. Six miles from Adirondack, Lake Colden, a mountain embosomed pond, 2,851 feet above the tide is reached, and from it foams and dashes the Opalescent river. Far up the gorge of the same name is seen Gray Peak, on which is Summit-Water, a clean pond 4,293 feet above the sea, whence flows the stream which afterwards becomes the mighty Hudson. One mile beyond Lake Colden is Avalanche Lake, surrounded by mountain peaks, and soon the slopes of Tahawus tower above us. A steady climb brings us to the summit, whence the descent may be made to Keeene on the east side of the mountains.

The Adirondack, or Indian Pass.

Continuing our journey northwardly from Adirondack, we pass through the celebrated Adirondack or Indian Pass, a great gorge between Mount Wallface on the left and Mount McIntire on the right. Its highest point is 2901 feet above the sea level, and for a mile old Wallface rises in an almost perpendicular precipice over 1300 feet high. The scene is one of wild and savage magnificence. The path is rugged and arduous, many times crossing the mountain torrent that makes its way through the pass. Great jagged masses of rock obstruct the way, and the climb upward to the ridge, or "divide," is enough to tax the stoutest legs and wear out the stoutest boots. The scene from the divide is thus graphically sketched by Stoddard:

"At last we near the summit and stand on Lookout Point; close by rises that grand wall a thousand feet up and extending three hundred feet below us; reaching out north and south, majestic, solemn and impressive in its nearness; a long line of great fragments have fallen, year by year, from the cliff above, and now lie at its foot; around on every side huge caverns yawn and mighty rocks rear their heads, where He who rules the earthquake cast them centuries ago. Along back, down the gorge we look, to where, five miles away, and 1300 feet below us, is Lake Henderson, a shining drop in the bottom of a great emerald bowl."

INDIAN PASS.

Here we leave the head waters of the Hudson behind, and descending by a route similar to that by which we have climbed from Adirondack, we follow the course of the Ausable, which takes its rise in this same notch, and passing through a dense forest, we reach, after five miles of tramping from the divide, the little hamlet of North Elba.

Near this point is the farm of old John Brown, of Ossawottomie, the hero of Kansas and of Harper's Ferry, and the huge bowlder by which he is buried, can be seen from the road. The farm is now the property of an association formed by Miss Kate Field. From this point the road runs northwest nine miles to Blood's, a tavern of some repute in the Saranac valley, and one mile further, reaches Martin's. East from North Elba two miles, is Scott's tavern, twelve miles further, over a most romantic route through a mountain pass, is Keene, and ten miles further is Elizabethtown, a beautiful village eight miles northwest of Westport, on the Boquet River, and encircled by mountain peaks.

Elizabethtown and its Objects of Interest.

This town, which lies in the centre of Essex County, is also reached by stage from Point of Rocks on the Ausable River. This is a favorite resort of quiet people, artists, ladies, and families who do not wish to get far from the base of home supplies. The hotels there are numerous, elegant, and well furnished, and naturally are well filled in the summer. There are two high summits on the west, of which the southermost (called the giant of the Valley) is one of the highest of the range. There is a perpendicular precipice on the north side of this, nearly 700 feet high. Five miles to the northwest is Hurricane Peak, a pyramid of naked rocks, graceful in shape, rising from a densely wooded base. Cobble Hill, one mile west of the village, has a precipice 200 feet high on the east side of it. The valley of the Boquet runs eight miles south-west from Elizabethtown. At the head of this valley, the Boquet has a fall of 100 feet, through a narrow gorge, over an inclined plane of rough and broken rocks. Black Pond is one mile long and half a mile wide; it is six miles south-east of the village, and well stored with fish. On the south-east of the town is a hill 200 feet high, covering 40 acres, supposed to be nearly ly a solid mass of iron ore. In the south-west the Ausable Ponds may be visited from Elizabethtown. These ponds, two in number, — the Upper and Lower, — are in the south part of the town of Keene, in the midst of scenery bold and wild. Hurricane Mountain and Skylight are easily reached from Ausable Ponds. Deep gorges, lovely little ponds, and wild cascades are found in the vicinity. About one-eighth of a mile west of the road leading from Keene Flats to Ausable Ponds, are the falls of the Ausable River, known as Russell's Falls. Here the water darts through a crooked passage one-third of a mile long, in which space it makes a descent of 150 feet, between rocky banks that rise to the height of 200 feet. Two miles farther up the Ausable, are similar falls, known as Beaver Meadow Falls. Rainbow Falls are one-eighth of a mile north-west of the Lower Ausable Pond, and have 125 feet of perpendicular descent. Roaring Brook Falls, four miles east of Rainow, bow, consist of two separate falls, — one over a vertical precipice into a deep gorge, the other 250 feet along a groove worn into the solid rock. Chapel Pond, the source of Roaring Brook, is about a mile east of Roaring Brook Falls, in a deep ravine between the Ausable and Boquet Rivers.

Lake Placid and the Wilmington Pass.

From North Elba nearly every sojourner pays a visit to Lake Placid, one of the loveliest of mountain lakes, lying high up among the peaks which circle it on every side. It lies two miles north of the village, and is five miles long by two wide. Three islands nearly divide it midway, and Mount Whiteface overlooks it on the northeast. Sugar Loaf towers on the west in dark, stern ridges, and tall peaks stand senti-

WHITE FACE MOUNTAIN FROM LAKE PLACID.

nels on the east. Near its southern shore stand the large boarding houses of Nash and Brewster, with accommodations for from 60 to 80 guests, and fishermen find in its clear depths ample rewards for "dropping a line."

The Wilmington Pass or Notch is the local title of the narrow valley through which the Ausable pours the waters of its west fork, which takes its rise in Lake Placid, and through which the carriage road from Wilmington to North Elba, twelve miles, has been constructed with immense labor and at great expense. The scenery in the pass is of wild and savage magnificence. The carriage-road is cut into the bank on the right side of the Ausable River; and above it tower, hundreds of feet, the rugged and perpendicular rocks. Across the river looms up old Whiteface, its cloud-capped peak 4200 feet above the sea, and its sides clothed with evergreen for a great part of its height. Midway of the notch are the celebrated Wilmington Falls, one hundred feet high, pouring over a precipice of eternal rock in a feathery cloud; with a roar like that of Niagara. A difficult and perilous climb down the walls of the gorge to the foot of the falls, is rewarded by a sight of their savage magnificence which is indescribably grand. A short distance below is the Flume, where the waters, compressed into a narrow space between high and smoothly worn walls of rock, rush with lightning rapidity down a steep incline; and all the way, the road, clinging to the mountain side, presents at each turn some new exhibition of Nature's power. Emerging from the pass, the peaceful, broad, and fertile valley of the Ausable stretches away for miles in the distance; and at our feet lies the little village of Wilmington. The Whiteface Mountain House, a cozy, homelike structure, with facilities for entertaining half a hundred guests, is the only hostelry and here nearly every one stops who proposes to ascend old Whiteface.

The Ascent of Whiteface.

From this House, a carriage road ascends the mountain on

its northeast side two miles; then a rugged bridle path is traversed on horseback four miles to the summit. Three quarters of a mile from the top is a rude hut, where quarters may be had for the night, if one desire to see the sunrise from the peak. The mountain is 4918 feet high, and is named from the fact that a landslide years ago laid bare its whitish-gray rocks near the summit. The ascent is attended with a good deal of difficulty, but is perfectly safe and often made by ladies, who find the Turkish costume specially convenient, as they have to follow the fashion of the men, and ride on both sides of the animal. The view from the summit is of surpassing grandeur, overlooking a vast territory broken by mountain peaks, among which shine the glassy surfaces of over sixty lakes and ponds. The Giant of the Valley, Marcy, Wallface, McIntire, Sugar Loaf and Seward are easily distinguished; Lake Placid nestles close to the mountain's foot, and afar off to the west can be seen the Saranac Lakes. To the east is the broad expanse of Lake Champlain, and beyond, in the dim distance, the outlines of the Green Mountains appear.

From Wilmington, we resume our journey northeast through Jay and Ausable Forks, to Ausable station, where we board the train for Plattsburgh. Or we may drive 24 miles over the plank-road to Port Kent, where we take the steamer for Plattsburgh.

CHAPTER IX.

Routes to Montreal.

FROM several points on our route thus far, the tourist who desires to visit Montreal, and the Canadian resorts, will find direct communication. From Boston there are several routes of nearly equal directness and advantages. We may journey to Rutland, as already described, and thence via Burlington. St. Albans, and St. John's to Montreal, or from Bellows Falls through White River Junction. Montpelier, Essex Junction, St. Albans, and St. John's to Montreal, or we may take the Boston, Lowell and Nashua Railroad, through Lowell, Nashua and Manchester to Concord, N. H., or the Boston and Maine Railroad *via* Lawrence and Manchester to Concod; and from the last named city, several divergent routes, all leading to Montreal, lie before us.

Via Lowell and Manchester.

Taking the morning train from the magnificent Passenger Station of the Boston, Lowell and Nashua Railroad, on Causeway street, we are soon rolling out of the city, across one of the many spile bridges that "spile" the beauty of the Charles

River near the city, while they serve to connect Boston with the continent north and to increase its material prosperity. We pass the Bunker Hill district, with its cosey and homelike State Prison at our right; Somerville, with its several small and scattered villages, and its imposing McLean Insane Hospital; Medford, famous for New England rum, for the Mystic Racing Park near the railroad, and for Tufts College, the handsome buildings of which are in plain sight, on the lofty hill overlooking the track on the left; Winchester, famous for its tanneries; East Woburn, North Woburn, Wilmington, Billerica and North Billerica, and cross the Concord river to Lowell. This great spindle city, 26 miles from Boston, is known wherever cotton cloth is worn. The vast power of the Merrimack river, which here descends 33 feet, over what were formerly known as Pawtucket Falls, is utilized by the canal, originally built for navigation, which connects with the Concord river below. Along this canal and upon the Merrimack and Concord rivers, stand long lines of huge factories, comprising the mills of the Lawrence, Tremont, Suffolk, Merrimack, Boott, Massachusetts, Middlesex, Prescott, Appleton, Hamilton and other corporations, besides the print and carpet factories. There are about 70 mills in all, employing about 1,000 women and half as many men, and some idea of the vast production may be gained from the fact that the Merrimack Mills alone turn out 12,000 miles of cotton cloth per annum. The operatives are mostly Irish, Nova Scotians and French Canadians. Besides the mills, a vast number of other profitable industries aid the wealth of the city. From Lowell, our route lies along the Merrimack river. Middlesex, at the head of the Old Middlesex canal, to Boston, completed in 1808 at a cost of $528,000, but abandoned since the era of railways, is passed; then North Chelmsford and Tyngsboro'; then we cross the New Hampshire line to the city of Nashua, 40 miles from Boston, which stands on both sides the Nashua river, a tributary of the Merrimack and the source of power for the various mills. Nashua is a city of over 10,000 inhabitants, which has grown since 1823, when the Nashua Manufacturing Company was chartered. Following the river Merrimack still, but over the line of the Concord Railroad, passing various rural stations, we reach Manchester, the largest city in the State, 57 miles from Boston, at the Amoskeag Falls of the Merrimack. The Blodget canal, around the falls, utilizing the power for the propulsion of various manufactories. The city has about 25,000 inhabitants and is very attractive in appearance, the streets being wide and handsomely shaded; several public squares are laid out and the houses are neat and many of them elegant. The falls, which have a descent of 47 feet, present a striking sight in high water. Lake Massabesic, four miles east of the city, is a favorate summer resort. Continuing our northward journey, we pass several little stations and enter the capitol of the State, the city of Concord, 75 miles from Boston, a beautiful town of some 15,000 inhabitants. The city stands on the west side of the river, to which the principal streets run parallel. The State Capitol fronts a small park off Main street, and is a stately structure of granite quarried near by. The City Hall, Court House, and State Insane Asylum, founded in 1842, are the principal public buildings. At Concord, the visitor who desires to make a stay of a few days, will find a pleasant stopping place at the Eagle Hotel,· John A. White, proprietor. This house has been enlarged, refitted and refurnished, and is a first-class establishment. Its location, opposite the Capitol, gives it the advantage in point of site over most other houses.

Via Lawrence and Manchester.

By the Boston and Maine line, leaving the commodious brick station in Haymarket square, Boston, our route is northward, across the Charles river, through one edge of the Bunker Hill District, Somerville, Malden, Melrose, Stoneham, Greenwood, Wakefield, Wilmington, Ballardvale, Andover, (the seat of Phillips Academy, the Abbot Female Seminary, and the famous Congregational Theological Seminary), and South Lawrence, across the Merrimack and into Lawrence, 25 miles from Boston, one of the great mill cities of the country, and one of the three capitals of Essex Country. It has about 30,000 inhabitants, and both by situation and the wise foresight displayed in its building and ornamentation, is one of the handsomest of manufacturing towns. In 1844, the Essex Company founded this place and built a massive stone dam, giving 28 feet fall, across the river. A canal, a mile long and 400 feet from the river, carries the water along the line of great mills, which stand on the strip of land between the river and the canal, which last thus separates the mills from the city. The corporation boarding houses are surrounded by a wide green, which gives them plenty of air and light, and a pleasant outlook. The streets are broad, handsomely shaded, and lined by many costly and elegant buildings, among which the City Hall, the Oliver High School, the County buildings, the Roman Catholic Church of the Immaculate Conception, and several other churches, are the most notable. The Common is a handsome large green, in the centre of the city, on which front several of the finest buildings, and which is shaded with beautiful trees. The principal factories are the Pemberton, Everett, Washington, Pacific, Atlantic, Arlington, Lawrence Woolen Company's and Russell Paper Company's Mills, employing 10,000 hands, and manufacturing millions of dollars' worth of goods annually. The valuation of Lawrence is about $20,000,000. The fearful accident in the Pemberton Mills, January 10, 1860, when the thin and insufficient walls were shaken down by the motion of the machinery, and the ruins took fire, burning to death many of the imprisoned operatives, is still fresh in the memory. By this terrible disaster 325 persons were killed and wounded.

From Lawrence, our route lies northwest, leaving the Boston and Maine for the Manchester and Lawrence Railroad, which conveys us in 70 minutes over the 26 miles of intervening distance. We pass by the way, Methuen, a flourishing village at the falls of the Spicket river, where hats, shoes, cottons, etc., are manufactured, and cross the New Hampshire line, into the quiet farming town of Salem; thence through Windham, Drury and Londonderry (settled by Irish Presbyterians and named after their former home; renowned for patriotism in the Revolution, and for the number of commanders in the continental army who were born here), we come to Manchester, whence our route to Concord is the same as already described.

From Concord to Montreal.

From Concord the principal lines are the Boston and Montreal Air Line, by the Boston, Concord and Montreal Railroad, through Weirs and Plymouth to Wells River, thence by Connecticut and Passumpsic Rivers Railway to Newport, Vt., thence by South Eastern Railroad to St. John's, thence by Grand Trunk Railway to Montreal; the Central Vermont line, via White River Junction, Montpelier and St. Albans; or by the Northern Railroad to White River Junction, Connecticut and Passumpsic Rivers Railroad to Wells River, Montpelier and Wells River Railroad to Montpelier, thence by the Central Vermont line; or by Boston, Concord and Montreal Railroad, through Weirs, Wells River, Littleton and Lancaster (with views of the White Mountains) to Northumberland, and thence by Grand Trunk Railway through Richmond Junction and St. Lambert to Montreal. This last is the longest and most tedious of the routes, but gives one an opportunity to see a good deal of the mountains and the country generally.

The Air Line Route.

By the first mentioned or Air Line route, we cross the Merrimack River, and pass through Canterbury, 82 miles from Boston, where is the seat of a flourishing Shaker community, whose good works (in the form of apple sauce and garden seeds), are known the country over; through Tilton (the seat of the New Hampshire Seminary and Female

College), to the shores of the beautiful Lake Winnepiseogee along which we skirt. Laconia has extensive manufactories of cloths, hosiery and railroad cars, and is extensively populated in the Summer by people from the cities, who find accommodations at the Willard House, and the various boarding houses and farm houses in the vicinity. Laconia is situated on the shores of Great Bay, also called Lake Winnisquam, a picturesque and extensive sheet of water, and is the point of departure for the ascent of Mount Belknap, four miles distant. This mountain, commanding from its summit a view of nearly the entire lake, is in sight from the car windows for several miles. Lake Village, on the shore of Sanbornton Bay, is a thriving place, with several lumber and paper mills. A small steamer runs in Summer between this "port" and Alton Bay. We now skirt the shores of Long Bay, and stop at Weirs, 105 miles from Boston, where we gain a magnificent view up the Lake, and where if we choose, we may take the steamer for Alton Bay, Woltboro and Centre Harbor.

A Tour of Lake Winnipiseogee.

Winnipiseogee, (spelt also Winnepesaukee), is an Indian word, variously interpreted to mean, "The Smile of the Great Spirit," and Pleasant Water in a High Place." Whichever be the correct version, either is applicable to this pure, clear, and wonderfully beautiful mountain lake. Edward Everett wrote of this trip to a friend :—

"I have been something of a traveller in our own country — though less than I could wish — and in Europe have seen all that is most attractive, from the Highlands of Scotland to the Golden Horn of Constantinople; from the summit of the Hartz Mountains to the Fountain of Vaucluse, but my eye has yet to rest on a lovelier scene than that which smiles around you as you sail from Weir's Landing to Centre Harbor."

Starr King wrote :—

"Looking up to the broken sides of the Ossipee Mountains that are rooted in the lake, over which huge shadows loiter;

or back to the twin Belknap Hills, which appeal to softer sensibilities with their verdured symmetry; or farther down upon the charming succession of mounds that hem the shores near Wolfboro'; or northward, where distant Chocorua lifts his bleached head, so tenderly touched now with gray and gold, to defy the hottest sunlight, as he has defied for ages the lightning and the storm, — does it not seem as though the passage in the Psalms is fulfilled before our eyes, — 'Out of the perfection of beauty, God hath shined.'"

This magnificent lake, the liquid gem of the Switzerland of America, is 25 miles long by 7 miles wide, at the widest, and contains 69 square miles. It is situated in Carroll and Belknap Counties, New Hampshire, and bordered by eight townships. Its shores are very irregular, expanding on every side into deeply indented bays, while some 300 islands break its surface. Long Island, far out in the lake, nearly opposite Weirs, and Diamond Island, several miles southeast, have small hotels, which are much visited by excursion parties. Taking the steamer Lady of the Lake, or Mount Washington, at Weirs, our course lies through a tortuous water path, between green and romantic islands, out upon the sparkling bosom of the lake. We turn to the left and gain a magnificent view, which fills every heart with rapture. Far away before us loom the Sandwich Mountains, in a long rank of sentinel peaks, the Ossipee Mountains furthest to our right, and grim Chocorua towering above and beyond them; then Whiteface, and to our left, Red Hill. We steam up the beautiful Northwest bay, between shores clothed in green, and fringed with trees, to Centre Harbor, a village of half a thousand permanent inhabitants, and as many more transient visitors in the season, where the Senter House, a large and pleasant hotel, and several smaller hostelries provide for the wants of the guests. From Centre Harbor there are several pleasant drives; to Red Hill, four miles of carriage road and two miles of bridle path to the summit, 2,000 feet high, whence the view is grand enough to repay ten times the exertion of climbing hither. Far away, on the northwestern horizon, rises the sharp, rocky cone of Chocorua; towards the north, the Sandwich range, with Bald Knob and Whiteface as its western outposts; to the north and northwest, the Squam Mountains; to the west Squam Lake, with its placid waters and its silvery beaches; to the southwest, 30 miles away, is Mount Kearsarge; to the south stretches the winding shore of the lake, with its beautiful islands, and beyond rise the twin peaks of Mount Belknap; while far away to the southwest, stretches one of the loveliest water panoramas on earth, backed by the domes of Copple Crown and Tumble-Down-Dick.

Down the Lake to Wolfboro'.

Leaving Centre Harbor for the trip down the lake, we steer southeast, leaving the Ossipee Mountains on our left, and threading our way through a labyrinth of islands for several miles. Emerging at length into the clear water of the centre of the lake, we cast our eyes to the eastward, where forty miles away Mount Washington's majestic peak, throned on the hills, towers above his satellites. Further on we gain shifting views of one after another of the magnificent peaks, while on the southern border of the lake, Mount Belknap rises before us as we advance. A sail of twenty miles brings us to Wolfboro, on the bay of the same name, a village of some thousands of inhabitants, with several hotels, banks and stores, lumber mills and a railroad station. Here the Wolfboro Branch of the Eastern Railroad has its terminus, and hence, if we choose, we can take the Pullman cars and speedily be transported through a charming country to Portsmouth, N.H., and thence to Boston, 106 miles, in less than five hours.

But we are not ready for so speedy a return, and can afford to enjoy for a time the delights of Wolfboro. The village is beautifully located on the lakeside slopes of two gently rising hills, separated by a millstream, and from almost every point fine views across the lake, with the majestic peaks of Belknap for a background, can be had. The Pavilion Hotel has a most sightly location, and during the season its broad verandahs and cool parlors are alive with city folk. Just by its side, embowered in trees, stands the Belvue, a quiet and homelike little house, much favored by modest and quiet people, and across the street, overlooking the lake, the steamboat landing and the railway station, and within a few rods of each, stands the Glendon House, a new and spacious hotel, with all the modern improvements. From either house boats are furnished for sailing and fishing on the lake, and teams for the many beautiful and romantic drives, as well as for the ascent of Copple Crown Mountain, 5 miles distant and 2100 feet high. From its summit the lake is visible for nearly its entire length, while Mounts Belknap, Ossipee, Chocorua, and Washington are the principal peaks in sight. In a clear day a view of the ocean is obtained. "Tumble-down Dick" is a neighboring and somewhat smaller mountain of singular formation, and is also often ascended. The Devil's Den, a narrow, black cave, among the rocks of a lofty hill a few miles away, is visited by active climbers, and there are many delightful short drives in the vicinity. But the great joys of the place are the sails and rows upon the lake, and picnic parties upon some of the many islands within easy distance. An apt poetical description, by Nathan D. Urner, of the pleasures of picnicing, tells the story of a day's sport in this line:

A leafy island, bowered by tall trees,
   A cove of silver, hushed from the far breakers;
A shallop, slanting shoreward with the breeze,
      Brimful of merrymakers.
The grating keel, the boat-stake nicely missed,
   Young fellows laughing at each other's error;
Hoops disarranged, curls flustered from their twist,
      The girls in pretty terror.

The shore: — bright ankles glancing up the sward,
   The strong arms and luncheon coming blithely after;
The nook selected, and the breathing hard,
      The jest and ringing laughter.

The rushing swing beneath the rustling limb,
   The youngsters pushing and the lasses soaring;
Romping and frolic by the fountain's brim,
      Croquet and battledoring.

Garlands of wild flowers shading many a brow,
　And garnered posies of wee blooms prized highly;
Strayed couples missed, and soft conjectures how
　They slipped away so slyly.

The snowy cloth upon the emerald sod,
　With platters few, but viands in profusion;
Ham and cold chicken — feasting for a god,
　With fear of no intrusion.

Cool bottles dripping from the icy spring,
　Old Sherry bubbling and bright Champagne popping;
Youth and high joyance, in a jocund ring,
　Both thirst and hunger stopping.

Blanche, with her beauty heightened by a blush,
　Which hints some secret of her recent ramble;
Maud, all the prettier for heated flush,
　Her skirt torn by a bramble.

Helen and Annie flirting with their beaux;
　Belle with her apron stuffed with dewy cresses;
Harry, the rascal, struggling for the rose
　In Edith's raven tresses.

Back to the cove: — a group of twelve or more,
　All homeward bent on tasks of love and duty;
A shallop, slowly gliding from the shore,
　Freighted with health and beauty.

Near Wolfboro, in a fine grove overlooking the Lake, the adventists have a noted camp-ground, and hither many of the sect yearly resort, to hear the prophecies expounded and to prepare for their speedy and final departure from this sublunary sphere. From Wolfboro, by the branch railroad to the Junction, 12 miles distant, and thence by the Conway Division of the Eastern Railroad, passengers can speedily reach North Conway and thence the White Mountains.

To Alton Bay and Back to Weirs.

Our route lies in the opposite direction. We resume our steamboat travel, steering nearly due South, into Alton Bay.

9*

This inlet, or frith, is of surpassing beauty, bordered by lofty bluffs, crowned by lofty trees, and the narrow water-way seeming, at times, to end abruptly before us, when rounding a bold point, we see it expanding in another direction. We follow it thus for four or five miles, to its head, where the railway station of the Boston and Maine, and the Bay View Hotel confront us. Alton Bay is one of the principal approaches to the lake, by the Boston and Maine, wh'ch forms a White Mountain route, by the aid of the steamer, to Centre Harbor, and stages thence to the mountain. Returning to Weirs, we resume our seats in the cars, for the prosecution of our journey northward. But before we leave the lovely lake, let us read John G. Whittier's beautiful description of an evening upon its fair waters.

Summer by the Lakeside.

Yon mountain's side is black with night,
    Whi e, broad-orbed, o'er its gleaming crown
The moon slow rounding into sight,
    On the hushed inland sea looks down.

How start to light the clustering isles,
    Each silver hemmed! How sharply show
The shadows of their rocky piles,
    And tree-tops in the waves below!

How far and strange the mountains seem,
    Dim-looming through the pale, still light!
The vague, vast grouping of a dream,
    They stretch into the solemn night.

Beneath, lake, wood, and peopled vale,
    Hushed by that presence, grand and grave,
Are silent, save the cricket's wail,
    And low response of leaf and wave.

Fair scenes! whereto the Day and Night
    Make rival love, I leave ye soon,
What time before the eastern light
    The pale ghost of the setting moon

Shall hide behind yon rocky spines,
   And the young Archer, Morn, shall break
His arrows on the mountain pines,
   And, golden-sandalled, walk the Lake!

Farewell! Around this smiling bay
   Gay-hearted Health, and Life in bloom,
With lighter steps than mine, may stray
   In radiant summers yet to come.

But none shall more regretful leave
   These waters and these hills than I;
Or, distant, fonder dream how eve
   Or dawn is painting wave and sky;

How rising moons shine sad and mild
   On wooded isle and silvering bay;
Or setting suns beyond the piled
   And purpled mountains lead the day;

Nor laughing girl, nor bearding boy,
   Nor full-pulsed manhood, lingering here,
Shall add to life's abounding joy,
   The charmed repose to suffering dear.

Still waits kind Nature to impart
   Her choicest gift to such as gain
An entrance to her loving heart
   Through the sharp discipline of pain.

Forever from the hand that takes
   One blessing from us others fall;
And, soon or late, our Father makes
   His perfect recompense to all!

O, watched by Silence and the Night,
   And folded in the strong embrace
Of the great mountains, with the light
   Of the sweet heavens upon thy face,

Lake of the Northlands! Keep thy dower
   Of beauty still, and while above
Thy solemn mountains speak of power,
   Be thou the mirror of God's love!

The Pemigewasset Valley, and Plymouth.

Next we pass Meredith, a pleasant village on the west shore of the lake, with which we now part company, and proceed northwest, through Ashland, the point of departure for the

THE PEMIGEWASSET HOUSE.

beautiful Squam Lake, three miles east, cross the Pemigewasset river near Bridgewater, and "round to at the wharf,—

no, beg pardon, the depot platform, — at Plymouth, 123 miles from Boston, just in time to hear the bell ringing for dinner at the Pemigewasset House close by. It is a singular and instructive coincidence, that at whatever time the train arrives, it is always dinner-time at the Pemigewasset, and as a fine table is set and thirty minutes are allowed for dinner, it is also a fortunate circumstance for travellers that it is so. The hotel has accommodations for about 300 guests, and is a favorite resort of tired pilgrims seeking rest among the lovely scenes in this vicinity.

Plymouth is one of the most beautiful, flourishing and attractive of mountain villages, and the inducements which it holds out to visitors are equalled by those of very few Summer resorts in the country. The ascent of Mount Prospect, with its splendid overlook, the drives and rambles along the romantic Pemigewasset, with its broad and fertile intervales, the trips to the mountains, to the fine Livermore Falls, 2 miles north of the village, to Squam Lake, six miles east, and to Newfound Lake, nine miles southwest, and the stage ride to the Profile House, in the Franconia Notch, White Mountains, 20 miles distant, are the principal features to be " taken in " during a stay here.

From Plymouth we continue our journey through Rumney and Wentworth to Warren, a place much frequented by visitors in Summer. The Moosilauke House furnishes comfortble accommodations, and the Summit House on the top of Mount Moosilauke, nine miles from the village by a good carriage road, (fare $4), also offers " the comforts of a home" to wayfarers, who choose to view the landscape o'er from the mountain top. The hight of Moosilauke is 4600 feet, and standing isolated as it does, the view from its summit is grand and extended. Hurricane Brook, near Warren, has many picturesque cascades, and there are many beautiful walks and drives in the vicinity. Passing through Haverhill, the county seat of Grafton county, we emerge from the mountain region upon the rich and fertile intervales of the Connecticut river, which we soon cross and draw up at the Wells River station, 165 miles from Boston.

The Grand Trunk Route.

Hence the Boston, Concord and Montreal Railroad, which runs parlor cars all the way to the White Mountains, continues northeasterly, recrossing the Connecticut to Northumberland, 219 miles from Boston, where it joins the Grand Trunk on its way from Portland to Montreal. From this junction, the Grand Trunk runs in a general northern direction, through the northwestern corner of New Hampshire, and the uninteresting and sparsely settled Canadian townships to Richmond Junction, whence branches diverge respectively to Quebec, Three Rivers and Montreal, the latter portion running almost due west, and traversing 76 miles in about three hours. Midway is St. Hyacinthe, a pleasant French Canadian city, with a fine cathedral, and a college said to be fully equal to any in Canada. The rest of the trip is through a fertile district of tillage land, occupied by inhabitants who preserve the manners, customs, religion and language of their French ancestors to a remarkable degree.

From Wells River, via Newport, Vt.

But the "Air Line" so-called, proceeds along the banks of the Connecticut river, almost due north, over the Connecticut and Passumpsic Rivers Railroad, to St. Johnsbury, 204 miles rom Boston. This is a busy and handsome town of some 5000 inhabitants, picturesquely located at the falls of the Passumpsic. The manufactory of Fairbanks scales is the principal industry, and gives employment to 500 or 600 men. Some 50,000 scales of various kinds are yearly made. There are also manufactories of mowing and threshing machines and other agricultural implements. St. Johnsbury is the shire town of Caledonia county, and has a fine court house on the hill, and in front a Soldiers' Monument — a pedestal surmounted by a statue of America, by Mead — and near at hand the Athenæum, with a library of some 10,000 volumes. Beyond St. Johnsbury our route takes us through the towns of Lyndon, in which are the Great Falls of the Passumpsic, and Burke, whence carriages may be taken for Willoughby Lake, six miles north. This lake is one of the most wonderful and interesting natural objects in the State or indeed in the country, being situated between two immense mountains, whose bases meet far below its waters. The Lake is six miles long two wide and of unknown depth, a 100-fathom line failing to reach bottom. Mount Hor, 1500 feet above the water and 2700 above the sea level, stands on the west, and Mount Pisgah, Willoughby or Annanance (as it is variously named) 2638 feet above the Lake, on the east shore. The latter mountain is ascended by a pleasant walk of two miles up its forest clothed slope, and a vast and panoramic view is obtained from the summit, stretching out over the White Mountains on the Southeast; Owl's Head and Jay Peak in Canada, on the Northwest; Mounts Mansfield, Camel's Hump and Killington on the Southwest; Lake Memphremagog to the North, while the Connecticut Valley stretches far away to the South. The western face of the mountain is a perpendicular cliff of granite, 600 feet high and two miles long. Rare flowers and plants are found at the "Flower Garden" at the foot of the cliff, and elsewhere on Mount Annanance. The Lake abounds in trout and muscalonge. The Willoughby Lake House furnishes accommodations for visitors, and hence carriages can be had for trips to other places of interests. Barton, 234 miles from Boston, is a place of considerable attraction to the tourist. Crystal Lake, with a house of the same name, is a lovely sheet of water near the railroad.

Lake Memphremagog and its Beauties.

We have now crossed the dividing ridge and are in the St. Lawrence water shed; a ride of fifteen miles more brings us to Newport, on the southern extremity of Lake Memphremagog, where we shall find it both pleasant and profitable to "tie up" for a time at the Mempremagog House.

This fine hotel (kept for several years by Mr. W. F. Bowman, a gentleman of long experience in the business, and one who never fails to enjoy the friendship and esteem of his guests, through his unvarying courtesy, efficiency and thoughtfulness), has just been greatly enlarged and improved, and now affords accommodation for several hundred guests, with all the luxuries and conveniencies of a palace. Its delightful location and healthful surroundings make it a most desirable summer residence. Water, gas, steam, bathing-rooms, billiard-halls, bowling alleys, a livery stable, pleasure boats, and a populous village, with everything that ministers to the traveller's occasional necessities, contribute to make the Memphremagog all that can be desired. One who has ever enjoyed the gorgeous sunset views from the broad piazzas, or sat on a moonlight evening while the band played, and watched the steamers and boats on the lake; one who has climbed Prospect Hill, roamed along the Clyde and Coventry Falls, within easy drive of the house, and then with sharpened appetite feasted on the speckled trout, the luscious berries, and other dainties of the mountain and the lake with which the tables here are loaded, — will need no urgency to bring him again to Lake Memphremagog. Jay Peak, in the towns of Jay and Westfield, thirteen miles west of Newport, is visited from here; and the ascent is effected by carriage road, — a magnificent view of the Green Mountains, the Lake, the White Mounttains, Lake Champlain, and the Adirondacks, repaying the tourist for the trip. Its height is 4018 feet. The magnificent body of water known as Lake Memphremagog, often likened by tourists to Loch Lomond, Lake Geneva, or Lake George, is 35 miles long and from 2 to 5 miles wide. Its rocky shores are indented with beautiful bays, while wooded headlands jut boldly out, and picturesque islands dot its surface here and there. Newport is a delightful village, upon a hillside sloping down to the clear water.

A Trip down the Lake.

This is one of the most delightful excursions that can be made. The "Lady of the Lake," a beautiful iron steamer, leaves the village every morning for Magog, a Canadian village at the northern outlet, and returns the same day. Indian Point, the Twin Sisters, Province Island, Tea Table Island, Fitch's Bay, and Whetstone Island are passed; and soon the steamer approaches "Owl's Head," a conical, symmetrical peak, rising 3,000 feet above the lake, whose waters lave its foot. A short sail past Round Island, a gracefully rounded and dense'y wooded islet, brings you to a landing almost at the foot of Owl's Head, where is located the Mountain House, a famous spot with those who delight in fishing; the deep, cool waters of the lake abounding in muscalonge and lake trout. The ascent of Owl's Head is made from this point, if one tarries long enough. Skinner's Island and Cave are near by, to the eastward of the Mountain House, and are famous as the haunt during the war of 1812 of Uriah Skinner, "the bold smuggler of Magog," of whom a poetic legend exists. In the cave, it is said, he took refuge from pursuit and there died. Continuing northward, Mount Elephantis (sugar Loaf) and the Hog's Back are seen; and we pass Long Island, on whose southern shore is the famous "Balance Rock," a huge mass of granite, poised on a narrow point at the water's edge. This island is the summer home of several wealthy Canadians, whose beautiful residences crown its wooded heights. Rounding the bold Gibraltar Point, Mount Orford comes in full view, — the loftiest peak of Lower Canada, rising 3300 feet, and distant five miles from the little hamlet of Magog, where the boat stops a short time. From Magog, John Norton's stage-line conveys the visitor who desires to Sherbrooke, on the Grand Trunk Railway, a ride of sixteen miles around the base of Mount Orford. At Sherbrooke, the Magog House, under the management of Mr. H. S Helpburn, will be found a very desirable resort. It is one of the best kept houses in Canada; and the finest fishing can be had in the vicinity.

The famous Lakes Massawippi and Meguntic are but a short distance from the house, and many visitors come hither to capture the trout, pickerel, bass, mullet, pike, muscalonge and other fine fish with which they swarm. From Sherbrooke, proceeding north to Richmond Junction and thence northeast to Point Levi on the shore of the St. Lawrence, but a short ferry ride separates us from Quebec; or, proceeding west as already described, we can reach Montreal. By far the shortest route, however, from here, is that by the Southeastern Railroad from Newport, via Richford and St. John's to Montreal. Our route lies in a northwesterly direction through the corner of Vermont into the realms of Queen Victoria. At Richford a connection is made with the Eastern Division of the Central Vermont for St. Albans, and at West Farnham, 65 miles from Newport, with the Northern Division of the same road, to Waterloo, 29 miles east. Fifteen miles further we come to St. John's, where we pass to the track of the Grand Trunk Railway, and cross the fertile township of La Prairie to St. Lambert, and thence by the magnificent Victoria Bridge, into Montreal. The cost of this gigantic structure was originally estimated at £1,450,000; but this sum has since been reduced, and the present calculation of its cost is about £1.250,000. In it 250,000 tons of stone and 7,500 tons of iron, have been used. The iron superstructure is supported by 24 piers and two abutments. The centre span is 330 feet: there are 12 spans each side of the centre, of 242 feet each. The extreme length, including abutments, is 7,000 feet. The height above summer water level in the centre opening is 60 feet, descending to either end at the rate of one in 130. The contents of the masonry is 3,000,000 cubic feet. The weight of iron in the tubes is 8,000 tons. The following are the dimensions of the tubes through which the trains pass in the middle span, viz., 22 feet high, 16 feet wide; at the extreme ends, 19 feet high, and 16 wide. The total length from the river bank is 10,284 feet, or about 50 yards less than two English miles. About ten minutes is spent in this dark passage, and then we emerge upon a lofty causeway, overlooking the majestic St. Lawrence and the city of Montreal, which we soon enter and in a few minutes draw up in the Bonaventure street station, a grim and dingy shed, unworthy the beautiful city and the wealthy and extensive Grand Trunk Railway.

The Central Vermont Route.

From Concord to Montreal one of the most direct and favorite routes, especially for passengers desiring to stop at any of the Vermont watering places by the way, is that by the Northern Railroad to White River Junction, and thence *via* Montpelier and St. Albans to St. John's. The road passes north from Concord, along the right bank of the Merrimack, by the manufacturing village of Fisherville, then crosses at the confluence of the Contoocock River to Dustin's Island, and thence to the left bank. On the island (which is noted as the spot where Hannah Dustin, who was captured by the Indians, at the sacking of Haverhill, Mass., slew a number of her savage guards and escaped) a granite monument to her was dedicated, with imposing ceremonies, in the Summer of 1874, on the spot of the slaughter. From this point we traverse the rich lands of Boscawen and Franklin, (the birthplace of Daniel Webster, in 1782, and now a thriving factory village, whence a branch runs 18 miles up the Pemigewasset Valley to Bristol) through the Andovers to Potter Place, whence stages run to Mount Kearsarge, four miles south. From its 2461 feet of elevation a fine view of the Green Mountains, Lake Winipiseogee and the White Mountains is obtained. The Kearsarge House, near the railway station, has good accommodations for the visitor. Thence we proceed to White River Junction, 144 miles from Boston, where we cross the Connecticut River on an open bridge, affording a fine view up and down, and draw up alongside the broad platform of the station. At this Junction the Central Vermont Railroad to St. Albans, Connecticut Valley Railroad from Bellows Falls and places south, the Passumpsic and Connecticut Rivers Railroad to Newport, Vt., and the road over which we have just passed, all centre. The Junction House at this place has long been a favorite with the travelling public. It offers special conveniences for those who are travelling between Canada and Boston or New York. Through trains arrive here at all hours of the day and night, making sometimes a long stop. This house, which is but a few steps from the station, and is every way a first-rate hotel, gives to all travellers needed rest and refreshment. There is also a large dining hall, well supplied, in the station, and as all trains stop long enough for a good honest dinner, the passenger is sure of good fare either at the station or hotel. Both are kept by the Messrs. A. T. & O. F. Barron, the famous hosts of the Crawford and Twin Mountain Houses, in the White Mountains. We now ascend the valley of the picturesque White River, for about 25 miles, to Roxbury, the summit of the pass, 1,000 feet above the sea level, by which we cross the Green Mountain chain. Here are quarries of splendid verd.antique marble, and just across a bridge, 400 feet long and 70 high, is the town of Northfield, renowned for slate quarries, and the Norwich Military Institute. Ten miles further, or 207 from Boston, is Montpelier Junction, whence a branch of a mile leads to the village, the Capital of Vermont Montpelier is a pleasant and attractive place, on the Winooski or Onion River, which makes its way hence to Lake Champlain, at Burlington. The village is situated on a sort of plateau surrounded by hills, and is neatly and compactly built. The State House, a fine edifice, of light granite, fronts on a beautiful common, and standing on a considerable elevation, is reached by terraces and steps of granite. It contains Larkin G. Mead's marble statue of Ethan Allen, and trophies of Vermont valor in the several wars of the nation, among them 2 cannon captured from the Hessians, at Bennington. The State House is surmounted with a dome 124 feet high, crowned with a marble statue of Ceres. The country about Montpelier abounds in pleasant drives. Beyond the

Junction, the main line passes through Middlesex to Waterbury, the point of departure for Mount Mansfield and Camel's Hump. The latter is eight miles south, and can be visited from Waterbury. On the way, 3 miles from the station, are the romantic Bolton Falls. Stowe, 10 miles north of Waterbury, is the objective point for visitors to Mount Mansfield, and a drive of 8 miles from Stowe brings one to the summit, 4348 feet above the sea level.

Stowe and Mount Mansfield.

Stowe is often called "The Saratoga of Vermont," from the number of summer visitors who flock hither, each season. Here is located the famous Mount Mansfield Hotel, accommodating 400 guests, with large airy rooms, in suites or private parlors, brilliantly lighted with gas, supplied with livery stables, bowling alleys, croquet grounds, cafe, theatre, and billiard halls. A telegraph office is near by; and the table is spread with the choicest delicacies of the season. A good road has been built to the top of the mountain, over which visitors can be taken with safety, without change of carriages. The summit is broken into 3 peaks, called the forehead, the nose, and the chin, from a fancied resemblance in their outline, to a human face turned upward. The Smuggler's Notch is a deep rocky pass between the Nose and Mt. Sterling, and was used in the war of 1812, for smuggling goods between Central Vermont and Canada. A small hotel stands in the Notch, near a great spring, the source of the Waterbury river. Many other places of interest are found near Stowe, and its popularity as a Summer resort is well deserved. Continuing our journey, the country growing more open and fertile, and the Winooski River being often in view, with its many falls and rapids, Essex junction is reached, 240 miles from Boston, whence a branch runs eight miles west, to Burlington, there connecting with the Rutland Division of the railroad. We follow the Winooski River closely, past the famous falls and gorges, and is a romantic bit of travel. The main line continues north from Essex, crossing the Lamoille river on a bridge 450 feet long, and stops at St. Albans, 264 miles from Boston. In Milton are the great falls of the Lamoille, 100 feet high, affording power for extensive lumber mils. Several fine views of the lake are gained during the last few m les.

St. Albans and its Advantages.

St. Albans is blessed with one of the most sightly locations in the world, and with the central offices and works of the Central Vermont Railroad, two advantages not lightly to be despised. It has also much history of warlike prowess to recount. It stands on a sort of table land three miles from Lake Champlain, which is spread out before the eye of any one who will drive about town and ascend Bellevue or Aldis Hills, and such a pro pect is rarely gained. Steamers cross the lake daily from St. Albans Bay, the nearest point on the shore, to Plattsburgh, N. Y. St. Albans has about 6000 inhabitants and one first-class hotel, the Weldon House. The great railway station of brick, with its commodious offices and its many tracks, is one of the finest in New England, and the great car shops near by furnish employment for several hundred men. This is the great butter and cheese market of New England, and St. Albans quotations rule the country in dairy products. In 1864, the famous raid of 22 armed Confederates from Canada occurred, resulting in the plundering of the banks of $208,000 in cash, the shooting of several citizens, one fatally, and a safe retreat into Canada. In June, 1866, a Fenian army assembled here, made a raid across the line, retired, and was disarmed by United States troops. In April, 1870, the same thing was repeated in every detail. Thus, though there has been little actual fighting here, St. Albans has been the scene of some very extensive wars on a small scale.

The Vermont Mineral Springs.

Ten miles from St. Albans, the Eastern Division of the

railroad — to Richford as previously noted — passes through Sheldon Springs, where are the famous Missisquoi, Sheldon, Vermont, Continental and Central Springs. The Missisquoi Falls, 119 feet high, are near the Sheldon Spring, which is a mile from the Missisquoi, and the Central Spring is two miles east. Near the Missisquoi stands the Missisquoi House, a large and first-class summer hotel; near the Sheldon Spring is Congress Hall, another fine house; and near the Central Spring are many hotels and boarding houses, going to make up Sheldon Village. The waters of these springs are celebrated for their cures of cutaneous disorders — dyspepsia and liver complaints, cancer, etc. They are extensively bottled, and also used for bathing.

On the main line north from St. Albans, a ride of nine miles past Swanton Junction and East Swanton, brings us to Highgate Springs, (where the Franklin House accommodates some 200 guests in the season). Two miles southeast, Highgate Falls on the Missisquoi river has another spring — the Champlain — and two hotels, the Champlain and Green Mountain. John G. Saxe, the poet, was born here, and this is the border town of Vermont, the train soon crossing the line into Her Majesty's Dominion of Canada, a fact which fails to impress us if we don't happen to know where it occurs. We now traverse several Canadian townships on the banks of the Richelieu, with the isolated peaks of Belœil and Rougemont visible on the right, and soon reach St. Alexandre, 307 miles from Boston, where we cross the Richelieu to St. Johns, whence we proceed to Montreal as already described.

Another route from St. Albans diverges from that last mentioned at Swanton Junction, whence we pass Alburgh Springs, 16 miles from St. Albans and 291 from Boston. Here are famous mineral waters used in cutaneous disorders, and a fine hotel, the Alburgh Springs House. There are fine drives on the lake shore, and excellent boating and fishing. Passing Alburgh and West Alburgh, our train crosses the outlet of Lake Champlain on a long trestle bridge, to Rouse's Point, N. Y., whence the Western Division of the Central Vermont, over which we have just come, continues northwest to Ogdensburgh on the St. Lawrence, 141 miles from St. Albans and 406 from Boston. The Champlain Division of the Grand Trunk Railway conveys us hence, along the left bank of the Richelieu, 23 miles to St. John's.

Yet another Route.

At White River Junction, if we choose, instead of proceeding northwest by the Central Vermont, we may follow the Connecticut and Passumpsic Rivers Railroad through Norwich, whence stages run to Hanover, about a mile southeast, the seat of Dartmouth College, to Wells River, 184 miles from Boston, whence the Montpelier and Wells River Railroad, opened in 1874, makes close connections for Montpelier, 38 miles distant. The line passes by a sweeping curve, up through Ryegate, Groton, Peacham, Marshfield and Plainfield to the capital, where we connect with the Central Vermont for St. Albans, by the route already described.

From Rutland to Montreal.

If it be desired to proceed direct from Rutland to Montreal, the Rutland Division of the Central Vermont, by which our route lay from Bellows Falls, will be adhered to, and we shall continue north to Burlington, 234 miles from Boston, 68 from Rutland. This route is very direct, and carries one through several places of considerable interest. At Sutherland Falls, a splendid view is presented from the car windows. A deep gorge, curtained by tall trees which almost obscure the vision of the falling waters, flashing through the rifts in the foliage as the branches are swayed in the wind, opens on the one side; while to the other it spreads out into a lovely intervale, with smooth, green meadows and smiling farms, while an amphitheatre of majestic mountains forms the background. As we journey onward, we see all along green and fertile meadows, mirror-like streams, and the grand array of mountains. Sixteen miles north of Rutland is Brandon. This town contains two singular caverns in limestone, eighteen feet square, entered by going down twenty feet from the surface. There is a marble quarry in the place; and minerals of different kinds abound. One of the notable curiosities and wonders of Brandon is to be found in the works of the Brandon Manufacturing Company. Howe's standard scales are made here, and have made their name familiar with all the dealers in goods sold by weight. At this place the Brandon House affords pleasant accommodations to those desiring to spend a few days in the vicinity and visit the numerous points of interest, or to those whose business brings them hither, of whom many every year avail themselves of its comfortable, homelike arrangements. At Pittsford, midway between Brandon and Rutland, is located a fine medicinal spring, its qualities being similar to those of the famous Clarendon Spring at Saratoga; and there is, besides, the best of trout-fishing in the immediate vicinity. The drives are delightful; and the marble quarries, like those of Rutland and Brandon, are of great interest to strangers. For their comfort the Otter Creek House affords a pleasant stopping-place.

Nine miles from Brandon by stage, and eight miles from Middlebury, is Lake Dunmore, a sheet of water nine miles long and three wide. It is surrounded by a pleasant variety of high bluffs, and green, sloping hillsides. The bracing mountain air and the fishing in the vicinity have made this quite a popular resort. The Lake Dunmore House is fitted up with special reference to the wants of visitors at the lake, and is well adapted to its purpose. Middlebury, fifteen miles north of Brandon, is on the Otter Creek, and has on every side most beautiful wild mountain scenery. White and variegated marble is found near by, and is exported in large quantities. This place is the seat of Middlebury College. Good fare and rest for travellers is furnished at the Addison House. From this centre the distance is short to Lakes

George, Dunmore and Champlain, Grand View Mountain, and Belden Falls. Soon the road approaches the shore of Lake Champlain, and we catch fleeting glimpses, through the openings in the trees, of its glittering, sheeny surface. Stretching away in the distance, its dancing waves, capped occasionally by a flitting sail, seem the embodiment of liquid life; while in the distance, looming hazy and only half substantial in their purple bloom, we see the peaks of the Adirondacks, far beyond the clear expanse of waters.

Vergennes, fourteen miles north of Middlebury, at the head of navigation on Otter Creek, is the oldest city of Vermont. There is an United States arsenal there; and much of the ship-building for Lake Champlain has been done at the place. Otter Creek is here 500 feet wide, and is navigable for the largest craft on the lake. At Vergennes, also, many visitors stop for a few days to enjoy the view of the falls, the sail down Otter Creek and upon Lake Champlain, and the delightful drives in the neighborhood. For all such tourists the Stevens House affords a desirable stopping-place. A steamer connects Vergennes with Port Henry on the west side of Lake Champlain, landing at Fort Cassin, Basin Harbor, and Westport, thus giving one of the direct routes to the Adirondack Mountains. The Otter Creek Falls, at Vergennes, are divided by an island, on both sides of which the water makes a descent of 35 feet. At Ferrisburg a dam, with its flashing sheet of water, forms the foreground of as delightful a scene as a painter would wish to put on canvas; while to the left, across the meadows, opens a picture of a quiet brook, and the neatly-kept grounds and cottage of some thrifty farmer. Near Shelburne we get a lovely view of Lake Champlain; and a little above, the cars pass very near to the water side, and we see the field of liquid azure stretching away from our very feet, with its emerald isles nestled cosily on the broad bosom of the lake. The purple peaks of the Adirondacks are in full view; and, with the soft haze which covers their rugged sides, they seem the enchanted mountains of our dreams. Soon the bright picture fades; the shores of the lake recede; and the train dashes through tracts of woodland obscuring the view, and giving only transient glimpses of the scenes we have so admired, and 21 miles from Vergennes, we emerge upon the very shore of the lake, between immense piles of lumber, skirt a long row of wharves and stop in the large station at Burlington. Thence our route to Essex Junction and St. Albans and so to Montreal has been already described.

From Plattsburgh to Montreal.

From Plattsburgh there are three routes to Montreal, neither of which requires long description. We may take the steamer to Rouse's Point and the railroad thence to St. John's as already described; or we may take the cars over the New York and Canada Railroad, to Mooer's, through an interesting country, newly cleared, flat and swampy, and thence to Rouse's Point. Or, if novelty be desired and time be no object, we may continue on the New York and Canada to Province Line, then change to the cars of the Grand Trunk which here connects, and solemnly meander northward, past the little Canadian villages — consisting chiefly of log cabins and tin roofed churches — of Hemmingford, La Pigeoniere, St. Remi and St. Isidore to the Indian Village of Caughnawaga on the south bank of the St. Lawrence. Here a squalid ferry-boat is taken for the transit across the river to Lachine, where we disembark, and in due time — it may be five minutes or it may be an hour — take another train for Montreal, 8 miles distant. The trip embraces only 63 miles from Plattsburgh, but it consumes four hours. At each of the stations the trains stops ten minutes or so, while the conductor and other train hands and such of the passengers as can talk the French Canadian *patois* "go ashore" and swap jokes with the loungers about the little depots. The loungers embrace pretty much the entire able-bodied population of each village, and they turn out *en masse* and stroll down to the station when the train comes, for their daily feast of gossip. This cannot be called an enlivening journey, but if one have a taste for the odd things of travel and plenty of time, it is worth trying. As "all roads lead to Rome" so each of the routes described will bring the wayfarer to Montreal, if that be the object of desire.

CHAPTER X.

Niagara Falls.

NIAGARA Falls can be easily reached from Saratoga, by resuming the cars of the Rensselaer and Saratoga Railroad, via Ballston to Schenectady, 23 miles southwest, and 252 miles from Boston. Schenectady is noted as the seat of Union College, which stands on the heights, overlooking the city. The buildings consist of two large halls, with a beautiful stone library between, and the University bears a high reputation. Schenectady stands on a broad plain, near the Mohawk, and is a city of 12,000 inhabitants; its iron works are the chief industries. Here we transfer ourselves to the dingy and comfortless wooden shed, dignified with the name of depot, and wait the arrival of the train over the New York Central Railroad, from Albany, bound for Niagara. For nearly 100 miles the road follows the valley of the Mohawk; and much of the time the river is in sight. At Fonda, 279 miles from Boston, a road diverges to Johnstown and Gloversville, 10 miles north, the principal manufactories of buckskin gloves and mittens in the country. At Little Falls, 309 miles from Boston, the Mohawk descends 45 feet, furnishing power for many factories. This town and Herkimer, seven miles further, are famous for the shipment of the rich cheese produced in the fertile country hereabout. Ilion, two miles beyond Herkimer, is a town of some 4000 inhabitants, where the famous Remington Arms Factory employs 1000 men. Twelve miles further the train crosses the Mohawk, on an iron bridge, and enters Utica, 330 miles from Boston, a city of 30,000 inhabitants, and the business centre of a rich and extensive farming country. The Erie Canal passes through Utica, and is joined by the Chenango Canal to Binghamton, 97 miles South. Utica is the seat of the Lunatic Asylum, and several well reputed educational institutions, and is a beautiful and pleasant place of residence. The Black River and Utica Railroad runs northwest to the St. Lawrence, at Clayton, and other railroad connections are made Southward to Norwich and Binghamton. By the Black River road, an excursion is made in 40 minutes to Trenton Falls, 17 miles North, on the West Canada Creek, a tributary of the Mohawk. The water descends 200 feet, by five cascades, plunging down into a deep chasm in the limestone rock, with perpendicular walls, from 70 to 200 feet high. The water is of a dark amber hue, and the various colors given out as it flashes over the rocks, and is lighted by the sun's rays, are very novel and beautiful. The Sherman, High, Milldam, Alhambra, and Prospect Falls are successively reached by ascending the canon from Moore's Trenton Falls House, and vary in height from a few feet to 80 (the High Falls.) The Alhambra, near the fall of the same name, is a broad stone platform, walled by the gorge. Rome, 344 miles from Boston, is quite a railroad centre for this region. It is a city of 11,000 inhabitants, with a United States Arsenal for its principal object of note, and the Rome, Watertown and Ogdensburgh Railroad, sweeping in a bold curve which follows the general outline of Lake Ontario, and the upper St. Lawrence, through Watertown, to Ogdensburgh and Potsdam, with branches to Sackett's Harbor, as its great artery of trade and travel. From Rome, we proceed southwest, by Verona, whose springs, saturated with muriate of soda and sulphuretted hydrogen gas, are strongly medicinal, and resemble the English Harrowgate springs, to Oneida, a village of 4,000 population, near the lake of the same, which we skirt on the South, at a few miles distance. Here the New York and Oswego Midland Railroad crosses the New York Central, running northwest to Oswego, on Lake Ontario, 57 miles. Near Oneida is the celebrated Free Love Commuity. Chittenango, 11 miles from Oneida, is the location of several medicinal springs, similar to the White Sulphur of Virginia. Kirkville and Manlius are passed, near which are other sulphuretted waters, and we arrive at Syracuse, 383 miles from Boston, the great salt manufactory of the country. Syracuse is a city of 55,000 inhabitants, located very near the geographical and railroad centre of the State, and hence is the meeting place of many conventions. Besides the New York Central, railroads run to Oswego and Sandy Creek on the north and northeast, to Burlington south, and to Norwich southeast. There are here many fine buildings, the Court House of Onondaga county, the Penitentiary, three Orphan Asylums, the State Idiot Asylum, City Hall, Syracuse University, two Convents and many churches being the chief.

The gray and imposing Renwick Castle near the University is seen over the trees, and produces a fine effect. The great curiosity, as well as the principal source of wealth of Syracuse, is the salt manufactory, which lies near Onondaga Lake, a little northwest of the city. Here great natural salt springs or wells yield in unlimited quantities a brine so strong that 35 gallons yield a bushel of salt. Some 5000 laborers, 40,000 solar vats and 20 kettles are employed, and 200,000 tons of coal annually burned. The production since 1797, when the State took control, has been some 230,000,000 bushels, and is now at the rate of 8,000,000 to 9,000,000 bushels. Horse cars convey the visitor to these works. Onondaga Lake, near which the Indian tribe of the same name used to have its fortress, is six miles long by one mile wide, and is traversed by small steamers and pleasure boats. From Syracuse to Rochester, the New York Central has two routes, — the old, through Auburn, Geneva and Canandaigua, passing near the northern ends of the Skaneateles, Owasco, Cayuga, Seneca and Canandaigua Lakes, which look on the map like so many sweet potatoes laid side by side, and at Canandaigua turns northwest and proceeds to Rochester, making the distance from Syracuse 102 miles; and the new or northern route, which passes in nearly a straight line through Port Byron, Lyons and Palmyra to Rochester, in 81 miles, or 464 from Boston. The through trains take the latter course. There is nothing of special interest to note until we reach Clyde, but at the latter village, 36 miles from Syracuse, we enter the paradise of mint juleps, the grand storehouse of peppermint in its crude form. Here are thousands of acres of mint on every side, one-third of all used in the country being grown hereabouts. It is cut and distilled, yielding 20 pounds or so of oil to the acre of herb. We now follow the Erie Canal, and catch frequent glimpses of the noble craft plowing the waters, under the impetus afforded by the gentle mule on the tow-path. To be the captain of one of these magnificent floating palaces, with its freight of grain and hops and happiness and hop-poles is indeed a proud ambition. Palmyra, 56 miles from Syracuse, is the place where Joseph Smith, the Mormon prophet, pretended to find the golden plates of the Book of Mormon.

Rochester and its Attractions.

Still following the Erie Canal, through a number of unimportant townships, we enter the beautiful outskirts of Rochester, cross the Genessee River just above its famous falls, and reach the railway station near the centre of the city. Rochester is a handsome city of 70,000 inhabitants, built on both sides of the Genesee River, seven miles from its mouth at Lake Ontario. It is the capitol of Monroe County, and the fine county buildings are located here. The City Hall, the University of Rochester, some 60 churches, several hospitals and asylums, numerous schools and business blocks, are also handsome and costly structures. The Powers Building is the most notable thing in the architectural line which Rochester has to show. It stands on the corner of Buffalo and State streets, and is probably without an equal in this country as a business structure. It fronts 175 feet on each street, and is seven stories high besides the basement, which is of New Hampshire granite, and a full story in height. The rest of the block is of Ohio stone and iron. The whole is surmounted by a square tower 30 by 24 feet, which rises five stories higher, the tiled "sky floor" being 163 feet above the street. The building is quadrangular in form, and tubular in construction, having an open area in the centre for the purpose of light. The ground floor contains a bank and fifteen stores, and the upper stories contain 220 rooms, used for purposes of business of all sorts, and for lodgings. Indeed, the block is a city in itself, and a person could live, carry on business, attend amusements, make calls, and view a good deal of the country without ever leaving the building. The halls are a l tiled with marble, the stairways are of iron, and all the floors are supported on iron girders and brick arches, the partitions are of brick and the window frames of iron. Each main wall rests on the solid bed rock. The grand staircase cost $20,000, and contains fifty tons of iron. The entire building is heated by steam. There are about 1000 tenants, and the building would hold 80 000 people on its ten acres of flooring. Two powerful elevators ascend the building, the car of one of which is the finest in the country. The view from the top of the tower, 400 feet above the level of the lake, is magnificent in its breadth and variety. In a clear day the Canadian shore is visible, and the commerce of the lake flitting by adds the element of life to the fair picture.

Another of the curiosities of Rochester is the great aqueduct by which the Erie Canal crosses the Genesee river. This was considered an impossible undertaking when the canal was projected, and its completion was deemed a wonderful triumph of engineering skill. The aqueduct is of cut stone supported on massive piers. The Genesee Valley Canal here joins the Erie, after descending 978 feet by 97 locks, in its 125 miles of length, from the Pennsylvania mountains. The Genesee River, which flows through the heart of the city, is bridged by Main street, which is built so solidly that one would never suspect a river flowed beneath, the buildings standing on stone arches through which the water passes. The Genesee and its famous falls have been the cause of Rochester's prosperity. The immense water power afforded by the falls has been utilized for the propulsion of many flowing mills, the wheat coming from the rich country near at hand. These falls are but a short distance from the center of the city, and are best seen from an enclosure, known as "Falls Field," with a small admission fee. The river here pours over a ledge of solid rock, 96 feet high, down into a walled gorge. The fall is very picturesque and only less majestic than Niagara. Here Sam Patch made his fatal leap. Nearly two miles below, the middle falls, 25 feet high, are reached by horse cars, and a little further on are the lower falls, 84 feet high, much the most picturesque of the three. Rochester is also famed for its nurseries of fruit trees and shrubs, for its flower seed establishments and for its beautiful streets, shaded by fine trees and bordered by handsome residences. From Rochester railroads diverge to Charlotte, seven miles North, where steamers touch on their way between Oswego and Toronto; to Buffalo, 69 miles West at the eastern extremity of Lake Erie, passing through the prosperous village of Batavia, the capital of Genesee county; to Cooming, 95 miles Southeast, the junction of the Erie Railway; Southwe-t to Caledonia, where the Canandaigua and Niagara Fall Railroad is joined, and West 77 miles to Niagara Falls.

Rochester to the Suspension Bridge.

From Rochester our route lies in a generally almost due westerly direction, passing through Medina (near which are the Oak Orchard Acid Springs), and Lockport, where the Erie Canal descends from the "Erie Level" to the "Genesee Level," 66 feet, by ten double locks of solid cut stone. These locks may be seen from the train, and give their name to the city. When viewed in the evening, with the long lines of lights on either side the canal and on the boats in the locks at the different levels, the scene is romantic and fairy-like. The canal also furnishes a large surplus of water which is used for driving flouring mills and factories. Here we cross the canal, with which we henceforth part company. From Lockport, a branch railroad runs southwest to Buffalo, 26 miles. Nineteen miles further brings us to Suspension Bridge, where the first view of the mighty falls is gained, and whence a ride of two miles along the eastern or right bank of the river, brings us to the village of Niagara Falls.

Niagara Village and its Hotels.

If we are to see the American side first, and the points accessible therefrom, we will continue on to the station above, in the village directly beside the Falls. There are several hotels near the station. The Monteagle House will compare favorably with any which the tourist will visit at any place. It has a fine location which affords a view of the two Suspension Bridges, two miles of the River, the entire Falls and the Whirlpool. The rooms are large, airy, and handsomely furnished, many of them overlooking the Rapids, Goat Island, and the Falls; the table is unexceptionable, being supplied with all the substantials and luxuries of the season; the attendance is excellent, the prices reasonable, and everything about the house is homelike and enjoyable. The proximity of the house to the mighty rapids (by many considered the most pleasing feature of Niagara), the views of Goat, Bath and other islands, and the ease with which one can visit every point of interest, have made this house justly a favorite. As for many years, the Monteagle House is under the management of Alexander and Terrill, which is sufficient guarantee of its continued popularity. The International, Spencer and other houses are in the village of Niagara Falls, and the Clifton House on the Canadian shore.

The Tour of the Islands.

The bridge to Bath Island has been mentioned. It is a substantial iron structure, crossing the river in the midst of the Rapids to Bath Island, on which there is a paper mill and several small buildings. A similar but smaller bridge crosses to Goat Island. The payment of 50 cents enables one to make the entire tour of the islands, or $1 pays for a season ticket. Goat Island is one of the wonders of Niagara. It would be a most charming and picturesque island anywhere, with its noble growth of forest trees, its sylvan dells, its fertile, sunny openings, and its delightful shade. But here, surrounded by the mighty river rushing onward to its stupendous plunge, the island seems to gain additional beauty; and new wonders develop themselves at every visit. At the upper end is the vast expanse of the Rapids; on either side the marvellous Falls, separated by the island itself, which, on its lower end, between the Falls, is a bare precipice of rock, rising sheer from the bed of the river over 150 feet in height. From Goat Island a small bridge crosses over the Central Fall, a lovely sheet of water, to Luna Island, — a tiny islet perched on the very verge of the precipice, and laved on either side by the Central and American Falls. At nearly the opposite extremity of Goat Island, a bridge is thrown across a small portion of the Rapids, to the first of three little islands, densely wooded, and known as the Three Sisters, other bridges connecting with the two remaining islets. From these a splendid view of the mighty rapids down to the edge of the great Canadian Fall, and the outline of its bold sweep, are gained. At the edge almost of this tremendous cataract another small foot-bridge is thrown across a slender channel of water, to a rocky prominence in the very edge of the Fall itself. Here stood Terrapin Tower, a circular structure of stone 45 feet high, famous for many years, from the top of which visitors could see the entire Falls, and even peer into the depths of the foaming abyss below. But in 1873, being deemed unsafe, it was blown up, and a new and more substantial structure is to take its place. Even the view of the Rapids from this point, or indeed from any point on the islands, would repay a visit. The river, with its mighty volume of water, pours over the rocky bottom, which has a rapid descent, — over 50 feet in three-quarters of a mile, — causing a succession of small cascades, boiling whirpools, and rushing channels.

The Biddle Stairs, and Cave of the Winds.

On the face of the vertical precipice between the falls, a wooden tower, encasing a spiral staircase, leads down to the bank of broken rock and *debris*, which has evidently fallen from the cliffs above, and forms a narrow dyke, sloping to the waters of the river. At the foot of these "Biddle Stairs" a person can emerge and walk for some distance beneath the cliff, almost to the foot of the great Canadian Fall, and, by a succession of wooden foot-bridges guarded by railings, into the "Cave of the Winds," as the space between the overhanging precipice and the sheet of water forming the Central Fall is called. In this cave the visitor, clad in rubber clothing, and wearing canvas overshoes to prevent slipping, is drenched with the spray which the currents of air, drawn in by the motion of the water, and meeting in the centre, are constantly whirling in every direction, while the tremendous roar of the vast body of water, beating on the rocks below, makes every

CENTRAL FALLS, (Cave of the Winds.)

other sound inaudible, and nearly deafens the tourist. From a visit to the " Cave of the Winds" a person can gain an idea of the immense volume of water constantly pouring over the Falls, this Central Fall being but a slender rivulet compared with the lofty American Fall, itself small in comparison with the great Canadian Fall. Near the Biddle Stairs is shown the spot where Sam Patch is said to have leaped from a projecting staging, down into the deep water below the Fall, and to have come out in safety. Goat Island is visited by thousands every year, the circuit of the island being made by carriages, and an opportunity given for the occupants to dismount at each point of interest.

Prospect Park and its Attractions.

Returning to the American shore from the tour of the islands, we proceed down the river bank a short distance to a lofty gateway inscribed "Prospect Park." Here 25 cents admits a foot passenger. Of this Prospect Park much nonsense has been written in the past year or two, about "fencing in the Falls," etc., and much denunciation of the owners of the land for their greed. No doubt the State of New York, years ago, ought to have reserved the lands lying along the Falls, as a part of the public domain, and made of them a free pleasure park; but the State did nothing of the kind. It sold grants of land to the settlers; and they have had to do the best they could. When Niagara first became a famous watering place, there were no improvements. Goat Island was almost inaccessible. There were no means of visiting the spots where now the most enjoyment is found. By degrees the islands have been opened to visitors, the ferry across the river established, the suspension bridges constructed. But Prospect Point—the projection of land directly abreast the American Fall, and from which one could toss a chip or even dip his hand into the very cataract — remained unimproved. It was a rough, rocky, scrubby cliff, covered with loose stones and gnarly trees, with no wall or railing even at the brink, to prevent the unwary from falling over; infested by peddlers, Indians, and vagabonds, and having no conveniences for the visitors. The land could not be made productive to the owners; for the hosts of visitors would constantly overrun it. So the owners associated themselves, and at the expense of several thousand dollars, have enclosed the Point, built a solid and safe wall along the edge of the precipice and on the side towards the Fall, so that visitors, even children, can sit with perfect safety directly over the rushing torrent, and gaze into its foamy depths; built a safe and rapid inclined railway (in a tunnel through the solid rock) to the water's edge below the fall, built summer houses, cleared and beautified the grounds, supplied them with seats, etc.

Across the River to the Canadian Fall.

A small fee pays for the descent of the inclined railway, and the ferry passage across the river to the Canadian shore. On this trip, which is made in a barge, one has a magnificent view of the whole of the falls at once, looking up from the river below them, which is here not broad but very deep, so deep in fact that the water appears a dark green. Reaching the Canadian shore, one can ascend a road to the bank above, or can scramble (if he be so minded, and have thick boots) over the stones along the water's edge to the foot of the great Canadian Fall, and ascend the stone stairs just below Table Rock. Then, having viewed the falls from the Canadian side, a short walk down the river brings us to the " New Suspension Bridge" (for carriages and foot passengers only) which crosses 1800 feet below the American Fall and by which we can return to this side. The towers of this bridge on the Canadian side are 120 feet high, and on the American side 106 feet.

Near Table Rock stands a small museum, from the roof of which a fine general overlook of the falls is gained, and where a collection of curiosities, insignificant enough beside the great curiosity of the world, the falls, is shown. Here guides and rubber suits may be procured for a trip under the edge of Horse-shoe Fall. Either tower of the Suspension Bridge may be ascended; that on the American side by stairs, that on the Canadian side by an elevator, and from either a majestic view of the Falls and the deep narrow gorge below is gained. Near the bridge, on the American side, is the beautiful Bridal Veil Fall, an artificial sheet of water, pouring over the perpendicular cliff, at the lower end of the hydraulic canal, which furnishes the power for some shops and for the hotels, and returns it to the river here.

The Great Suspension Bridge, and the View Thence.

But the Suspension Bridge, known the world over as one of the greatest achievements of engineering skill, is two miles below the Falls. Its length is 800 feet, and its height above the water 268 feet. The towers are 66 feet high; and each of the four main cables supporting the bridge is nine inches in diameter, and composed of 800 wires. There is a carriage and foot way 28 feet below the railroad track. Its cost was $500,000. Over it run, or rather crawl the trains of the Great Western Railway. for Hamilton and the far West. One mile below the Suspension Bridge the river widens, and gives a sudden turn, so that the waters are forced along in an immense seething, heaving whirlpool. A mile below the Whirlpool is the Devil's Hole, 150 feet deep, and two acres in extent. The carriage road runs right up to the margin of the abyss, so that without leaving a carriage one may look down into it. From the Suspension Bridge a splendid view of the entire Falls is gained. It is like a panorama or a bird's-eye view, so complete, yet so reduced by distance; and many think it the finest view that can be anywhere gained of the great wonder, Niagara. At the Suspension Bridge, the waters of the river are compressed into a narrow gorge, with high perpendicular cliffs for banks. From their top one can look down 240 feet to the surface of the water. The bottom of the stream is probably as much farther down; at any rate, the water from its immense depth, looks as darkly green as the ocean itself. Just below, the river narrows to 400 feet, one-tenth its width at the Falls, and here the water, from being so closely compressed, rushes through the gorge in the most terrible rapids, which toss and heave white masses in the center of the river to the height of 30 or 40 feet. A vertical railway leads to the water's edge at these famous Whirlpool Rapids. On the American side, at the Suspension Bridge, is Niagara City, (mostly on paper at present) nicely laid out in squares, with a large and pleasant hotel, the Monteagle. On the Canadian side is the village of Clifton.

Niagara Swindles, So-called.

Much denunciation has been wasted on Niagara hotel-keepers, Niagara hackmen, and Niagara swindles generally; and it is mainly based on ignorance or injustice. The hotel charges are no higher than at any other prominent summer resort; there are no more "extras;" the fees for seeing the wonders are — as we have seen — very reasonable, considering the attractions; and the charges of the hack-drivers are quite moderate, if one be not over-flush with his money at the outset. The hackmen are all licensed by the corporation of the village; and any complaint of over-charge or incivility, will secure the revocation of a license. For $2 (and the gate fees) a good carriage can be had to take one around the islands and through Prospect Park; for $5 two persons can ride all the forenoon. Of course there are petty swindles in the shops for the sale of "Table-rock jewelry," Indian bead-work, feather fans, etc.; but no one is obliged to buy them. And excellent stereoscopic views of the Falls, which are decidedy the prettiest and most useful mementoes to bring away, can be bought as cheaply as the same class of goods in New York or Boston. Of course the class of persons who go to Niagara merely to *say* that they have been there, and have seen all the sights, can also generally have it to say that they were outrageously swindled while there; but a sensible person, who goes to see the great wonders, fully and judiciously, can get a dollar's worth for every dollar he expends, as well as in New York or Boston. The neighborhood is full of historic associations connected with the late war with Great Britain. Fort Erie, Chippewa, Lundy's Lane, and many other scenes of hard-fought battles, are near. Lewiston and Queenstown are on opposite sides of the river, seven miles below the Falls, at the head of navigation on Lake Ontario.

Daniel Webster's Famous Description.

Daniel Webster's description of Niagara Falls, written in 1825, and found in vol. ii., p. 385, of his correspondence, has been often quoted; and some passages from it are worthy of reproduction here, though many things which he describes are changed since 1825 : —

"Lake Erie is 330 feet higher than Lake Ontario; but, in decending the river from Lake Erie, one perceives no very great descent, although the current is all the way rapid, till we get nearly down to the Falls. A little below the village of Black Rock, perhaps about five miles from Lake Erie, the river divides into two channels, forming a large island in the centre called Grand Isle, about 12 miles long, and in some places six or seven broad. This island terminates, and the two channels unite again, just at the head of what is called the Rapids, a mile or a mile and a half above the great Falls. These rapids are a succession of cascades spreading over the whole river, of different and various heights and appearances, rendering the whole breadth of the stream, (which is here not less than two miles) white with foam. They would form a fine object, if there were nothing near to call the attention another way. Midway of these rapids is Goat Island, which divides the river into two unequal parts, about one-third in breadth being on the eastern or American side, and two-thirds on the British. The island runs down to the very brink of the Falls, and there terminates in a perpendicular precipice (a wall of rock), which is part of the same great declivity over which the river pours. This island thus divides the river, so that it falls over the precipice in two sheets. The length of the fall on the American side is estimated at 380 yards; the distance across the end of Goat Island 330 yards; the length of the fall on the British side 700 yards. The fall is thought to be the highest on the American side, being there 165 feet, and on the British side 150. Vastly the greatest portion of water (three-fourths, or even more) runs on the British side. As you descend the river from Lake Erie and approach the Falls, the river seems to fall away from your feet, and to pitch right down into the earth. Many miles before you reach the Falls you see the mist or spray rising like a cloud; but this does not seem to be rising from the earth into the air as much as from the centre of the earth to the surface: it appears to be coming from the ground. From the bottom of the Falls to Lewiston, seven miles, the whole channel of the river is one great trough, 100 or 150 feet deep, with sides of perpendicular rock. This has given currency to the opinion that the Falls were once seven miles lower down than they now are, and that the force of the water in time has worn away the rocks, and forced the Falls back to their present position. As we stood to-day at noon, on the projecting point at Table Rock, we looked over into the abyss; and, far beneath our feet, arched over this tremendous aggregate of water, we saw a perfect and radiant rainbow. This ornament of heaven does not seem out of place in being half way up the sheet of the glorious cataract; it looked as if the skies themselves paid homage to this stupendous work of nature. From Table Rock, or a little further down, a winding staircase is constructed, down which we descend from the level of Table Rock, 95 feet. This brings us to the bottom of the perpendicular rock; and from this place we descend 50 or 60 feet further, over large fragments of rock, and other substance, down to the edge of the river. If at the bottom of the staircase (instead of descending further) we choose to turn to the right and go up the stream, keeping close at the foot of Table Rock or the perpendicular bank, we soon get to the foot of the fall, and approach the end of the falling mass. It is easy to get in behind for a little distance between the falling water and the rock over which it is precipitated. This cannot be done, however, without being entirely wet. From within this cavern there issues a wind, occasionally very strong, and bringing with it such showers and torrents of spray, that we are soon as wet as if we had come over the Falls with the water. As near to the fall in this place as you can well come, is perhaps the spot on which the mind is most deeply impressed with the whole scene. Over our heads hangs a fearful rock, projecting like an unsupported piazza. Before us is a hurly-burly of waters too deep to be fathomed, too irregular to be described, shrouded in too much mist to be clearly seen; water, vapor, foam, and atmosphere are all mixed up together is sublime confusion. By our side, down comes this world of green and white waters, and pours into the invisible abyss. A steady, unvarying, low-toned roar thunders incessantly upon our ears. As we look up we think some sudden disaster has opened the seas, and that all their floods are coming down upon us at once; but we soon recollect that what we see is not a sudden or violent exhibition, but the permanent and uniform character of the object which we contemplate. There the grand spectacle has stood for centuries — from the creation, as far as we know, without change From the beginning it has shaken as it now does the earth and the air; and its unvarying thunder existed before there were human ears to hear it. Reflections like these on the duration and permanency of this grand object naturally arise, and contribute much to the deep feeling which the whole scene produces. We cannot help being struck with a sense of the insignificance of man and all his works, compared with what is before us."

Excursions from Niagara.

From Niagara, trips may be made by carriage to the battlefields of Lundy's Lane and Chippewa, and to the Burning Spring, (which is kept in constant ebullition by a stream of sulphuretted hydrogen gas, which ignites and burns with an intermittent blue flame and the odor of aged eggs), all on the Canada side. By rail, one may proceed to the Suspension Bridge, two miles, or to Lewiston seven miles down the river on the American shore, and back, a rapid and inexpensive excursion.

Lake Ontario and the Upper St. Lawrence.

A most delightful trip is that from Niagara down Lake Ontario and the Upper St. Lawrence, to Montreal. Two routes may be chosen — one by rail to Kingston or Prescott, thence by steamer down the St. Lawrence; the other by boat across Lake Ontario, and down the river. Those who desire, or are compelled by urgency of time, can make an all-rail trip from Toronto to Montreal; but they will lose the charming scenery of the St. Lawrence and the Thousand Islands. In either case we visit Toronto; and to get there take a seven-mile railroad ride down the Niagara River, overlooking the stream much of the way, to Lewiston, which is situated at the head of navigation on the lower Niagara, and is a pleasant, well built village. Queenston is a village of about 200 inhabitants on the Canadian side, nearly opposite Lewiston, and was the scene of a battle in the war of 1812. Near this point the river becomes more tranquil, the shores less broken and wild, and the scenery changes from rugged grandeur to beauty. On Queenston Heights, the scene of the battle, stands Brock's monument, erected in honor of the British general who so gallantly defended the place.

Taking the little steamer "City of Toronto" at Lewiston, we are soon steaming down the Niagara river, on both banks of which are points of historic interest, dating from the days of the "Old French War," as well as the last war between the United States and Great Britain. Fort Niagara stands at the river's mouth, on the American side. There are many interesting associations connected with the spot, as, during the earliest part of the past century, it was a scene of many severe conflicts between the whites and the Indians, and subsequently between the English and the French. The village adjacent to the Fort is called Youngstown, in honor of its founder, the late John Young, Esq. Niagara is one of the oldest towns in Upper Canada, and was formerly the capital of the province.

Across the Lake to Toronto.

Leaving Niagara, we steam across the western end of Lake Ontario, and soon arrive at Toronto, the capital city of Upper Canada, which is situated on an arm of Lake Ontario, 36 miles from the mouth of Niagara river. Toronto Bay is a beautiful inlet separated from the main body of Lake Ontario, except at its entrance, by a long, narrow, sandy beach. The southwestern extremity is called Gibraltar Point. The population in 1817 was 1,200, but at the present time it amounts to about 60,000. Among the principal buildings of Toronto are

Trinity College, University of Toronto, and St. James Cathedral. One of the ecclesiastical edifices deserves especial notice, — the Church of the Holy Trinity, a handsome structure, erected by a donation of £5,000 from some liberal person from England, on condition that the whole of the seats should be free. The Elgin Association, for improving the moral and religious condition of the colored population, is among the most useful institutions of the place. That stupendous undertaking, the Grand Trunk Railway of Canada, passes through Toronto, and promises a splendid future for Toronto and its sister cities.

Down Lake Ontario to the St. Lawrence.

From Toronto, where we transfer ourselves to a much larger and finer steamer, — the "Corinthian," "Corsican," "Spartan," "Algerian," or "Bohemian," of the Canadian Transportation Company, — we proceed eastward, straight down Lake Ontario, keeping within a few miles of the northern shore. On this side, Port Hope, a pretty town containing about 2200 inhabitants, is located in the valley of a small stream emptying into the lake, with a fine range of hills rising to the westward. Coburg lies seven miles below Port Hope. It contains 4000 inhabitants, seven churches, two banks, and the largest cloth-factory in the province. It is also the seat of Victoria College and a theological institute. Kingston, the original capital of Canada, is at the mouth of the Cataraqui River, and just at the foot of Lake Ontario, whence runs the St. Lawrence. As early as 1672, the French under De Courcelles began a settlement here, and built a fort, which was named Fort Frontenac, in honor of the French count of that name. In 1762 the English took possession, and called the place Kingston. It is one of the important military posts of Canada, and has about 11,000 inhabitants. The harbor is very fine. The land projects out on the east side of the bay, forming Point Frederic or Navy Point, east of which is a deep basin called Haldimand Cove, where are found the royal dock yard, and much of the shipping of the navy. The city is built chiefly of blue limestone; and wells of mineral water have been found by boring to different depths, from 75 to 1145 feet. Among the noticeable buildings here, are the Roman Catholic Cathedral, the buildings of Queen's College (Presbyterian), Regispole's College (Roman Catholic), and the Provincial Penitentiary. The extremity of the Rideau Canal, which connects Lake Ontario with the Rideau River — one of the tributaries of the Ottawa — is near Kingston, and adds much to the business of the place. On the American side of the lake are Charlotte, Oswego and Sackett's Harbor.

The Thousand Islands.

About six miles below Kingston the river widens, and embosoms the loveliest group of islands imaginable, — the far-famed Thousand Islands. They are in an expansion of the St. Lawrence, at the outlet of Lake Ontario; and the broad river in which they lie partakes so much of the character of a lake, that it is often called "The Lake of the Thousand Isles." For 40 miles down the river this beautiful scene continues, the boat which leaves Kingston at early dawn gliding among no less than 1800 of these "emerald gems in the ring of the wave" of all sizes, from the islet a few yards square to miles in length, and covered with a heavy growth of trees. This group is constantly attracting the attention of sportsmen and pleasure-seekers. Fish so large as to make angling tiresome, and wild-fowl of all kinds, are everywhere abundant. President Grant has been a guest here of Mr. George M. Pullman, President of the Pullman Palace Car Company, who owns a villa on one of these islands. These islands, too, have been the scene of most exciting romance. From their great number, and the labyrinth-like channels among them, they afforded an admirable retreat for the insurgents in the last Canadian insurrection, and for the American sympthizers with them. Among these was one man, who from his daring and ability, became an object of anxious pursuit to the Canadian authorities; and he found a safe asylum in these watery intricacies, through the devotedness and courage of his daughter, whose inimitable management of her canoe was such that, through hosts of pursurers, she baffled their efforts at capture, while she supplied him with provisions in these solitary retreats, rowing him from one place of concealment to another, under shadow of the night. But, in truth, all the islands, which are so numerously studded through the whole chain of those magnificent lakes, abound with materials for romance and poetry. Opposite the Thousand Isles, on the American side of the river, is Clayton, well known as a lumber station. Here the rafts are made up for their long voyage down the St. Lawrence, and look like floating villages with the huts that are built on them for the protection of the raftsmen. Alexandria Bay is the next port after leaving Clayton. It is built upon a massive pile of rocks; and its situation is romantic and highly picturesque. It is a place of resort for sportmen, and during the past two or three seasons has become a popular and fashionable watering-place. Here the Thousand Island House, a fine hotel, built in 1873, furnishes palatial accommodations for 600 guests. The verandah connecting with the long hall of the first floor, gives a promenade 624 feet in length, the verandah portion being 374 feet and the hall 250 feet. The whole house is supplied with water, and lighted with gas. The view over the islands from the lofty tower is exceedingly fine. Alexandria Bay is 30 miles from Cape Vincent, and 36 miles from Ogdensburgh, both northern termini of the Rome, Watertown and Ogdensburgh Railroad. From both places, steamers ply to the bay. Some two or three miles below the village, is a position from whence 100 islands can be seen at one view. This place also is celebrated for its fishing and shooting. The beauty of the islands in this vicinity, for several miles up and down the river, can hardly be imagined without a personal visit. Here many of those splendid fish, the muscalonge, are killed: they are of large size, many of them weighing 40 or even 70 pounds each. They are taken with trolling lines; and it requires a skilful angler to land one safely. Sportsmen consider the taking of these fish equal to salmon-fishing. During the past few seasons many of these beautiful islands have been bought by wealthy people for Summer residences. Hart's island, directly opposite, and very near to the village, is said to be the spot where Thomas Moore, the Irish poet, wrote early in the century his famous

Canadian Boat Song.

Faintly, as tolls the evening chime,
Our voices keep tune, and our oars keep time;
Soon as the woods on shore look dim,
We'll sing at St. Ann's our parting hymn.
Row, brothers, row, the stream runs fast,
The rapids are near and the daylight's past.

Why should we yet our sails unfurl?
There is not a breath the blue wave to curl!
But, when the wind blows off the shore,
Oh! sweetly we'll rest on our weary oar.
Blow, breezes, blow, the stream runs fast,
The rapids are near and the daylight's past!

Utawa's tide! this trembling moon,
Shall see us float over thy surges soon:
Saint of this green isle! hear our prayers,
Oh! grant us cool heavens and favoring airs.
Blow, breezes, blow, the stream runs fast,
The rapids are near and the daylight's past!

On the Canada side, fifteen miles below Alexandria, is Brockville, one of the most attractive towns on the river, named in honor of Gen. Brock, who fell at Queenston in 1812. Here is the Junction of the Grand Trunk Railroad with Brockville and Ottawa Railroad, which extends northward to the Ottawa River.

Ogdensburg and its Railway Facilities.

On the American side of the river is Ogdensburg, a town of

about 9,000 population. This is the western terminus of the Ogdensburg and Lake Champlain Railroad (now under control of the Central Vermont), which connects Ogdensburg with Rouse's Point on Lake Champlain, and so opens the route to Boston and New York. The Central Vermont Company has here a freight and passenger station 305 feet by 84, and numerous other buildings for business on a grand scale. The extensive elevators of the Central Vermont line are located here, at which vessels laden with grain on the lakes discharge their cargoes. Opposite Ogdensburg is Prescott; and a mile below is Windmill Point, where the ruins of an old windmill are seen, in which Von Schultz took refuge with the Polish patriots in 1837. Five miles below, at the first rapids of the St. Lawrence, is Chimney Island, where the remains of an old French fortification are seen.

Excursion to Ottawa.

At Prescott, passengers can take the cars for Ottawa, and then descend the Ottawa river to Montreal. The distance from Prescott to Ottawa, over the St. Lawrence and Ottawa Railroad is 51 miles. Ottawa is the capital of the new dominion of Canada, and is situated on the Ottawa river, a stream 800 miles long, which enters the St. Lawrence on both sides of the island of Montreal, 130 miles below the city of Ottawa. The city is divided into three parts — Lower, Central and Upper Town. The Government Buildings, when completed, will be among the finest on the American continent. These buildings, with the government offices and Queen's Printing-house, occupy three sides of a square on the summit of Barrack Hill, which rises almost perpendicularly from the river to the height of 350 feet. Rideau Falls, in the eastern part of the city, two in number, are very attractive, but are far surpassed by the Chaudiere Falls in the western portion of the city. They are 40 feet high and over 200 feet wide. A suspension bridge just below the falls crosses the river, and gives a splendid view of the falls, the caldron below them, and the rapids. The lumber shoots which are built here for running down the lumber, and save it from breaking to pieces in going over the falls, are objects of exciting interest. The passage may be made from Ot'awa to Montreal by steamer down the Ottawa river. Picturesque and thickly wooded banks rise on each side much of the way. Two miles below Ottawa is the mouth of the Gatineau, a stream more than 400 miles long, which drains a vast unexplored region.

The Rapids of the St. Lawrence.

But a most exciting as well as one of the most delightful portions of our trip, is at hand, — the passage of the rapids of the St. Lawrence. At Chimney Island, previously mentioned, the first of these rapids, and one of the smallest and mildest, — the Galop Rapid — is reached. Next comes the Long Sault, a continuous rapid of nine miles, divided in the centre by an island. The usual passage for steamers is on the south side. The passage is very narrow; and such is the velocity of the current, that a raft it is said, will drift the nine miles in forty minutes. This is the most exciting part of the whole passage of the St. Lawrence. The rapids of the " Long Sault." rush along at the rate of something like 20 miles an hour. When the vessel enters within their influence, the steam is shut off, and she is carried onwards by the force of the stream alone. The surging waters present all the angry appearance of the ocean in a storm; the noble boat strains and labors; but, unlike the ordinary pitching and tossing at sea, this going down hill by water produces a highly novel sensation, and is, in fact, a service of some danger, the imminence of which is enhanced to the imagination by the tremendous roar of the headlong boiling current. Great nerve and force and precision are here required in piloting, so as the keep the vessel's head straight with the course of the rapid; for if she diverges in the least, presenting her side to the current, or " broached to," as the nautical phrase is, she would be instantly run aground. Hence the necessity of enormous power over her rudder; and for this purpose the mode of steering affords great facility; for the wheel that governs the rudder is placed ahead, and by means of chain and pulley sways it. But, in descending the rapids, a tiller is placed astern to the rudder itself, so that the tiller can be manned as well as the wheel. Some idea may be entertained of the peril of descending a rapid, when it requires four men at the wheel and two at the tiller, to insure safe steering. Here is the region of the daring raftsmen, at whose hands are demanded infinite courage and skill. There is, however, but little danger to life, as it frequently happens that a steamer strikes and sinks; but a few minutes puts her safely in shoal water. The Canadian Navigation Company has never lost any lives by accidents of this kind in descending the rapids. Of course it is impossible for steamers to ascend these rapids; so canals are constructed around them, with locks, by which the boats are enabled to make the return passage. The splendid boats of the Canadian Navigation Company leave the foot of Lake Ontario in the morning, and reach Montreal at night. The Government is about to deepen the channel through all the rapids to 10 feet. Cornwall, at the lower end of the rapids, is near the boundary line between the United States and Canada. St. Regis is an old Indian village, and lies a little below Cornwall, on the opposite side of the river. It contains a Catholic Church, which was built about the year 1700. While the building was in progress, the Indians were told by their priest that a bell was indispensable in their house of worship, and they were ordered to collect furs sufficient to purchase one. The furs were collected, the money was sent to France, and the bell was bought and shipped for Canada. But the vessel which contained it was captured by an English cruiser, and taken into Salem, Mass. The bell was afterwards purchased for the church at Deerfield. The priest of St. Regis having heard of its destination, excited the Indians to a general crusade, for its recovery. They joined the expedition fitted out by the governor, against the New England Colonists, and proceeded through the then long, trackless wilderness, to Deerfield, which they attacked in the night. The inhabitants unsuspicious of danger, were aroused from sleep only to meet the tomahawk and scalping-knife of the savage. Forty-seven were killed, and 112 taken captive; among whom were Mr. Williams, the pastor, and his family. Mrs. Williams being feeble at the time and not able to travel with her husband and family, was killed by the Indians. Mr. Williams and part of his surviving family afterwards returned to Deerfield, but the others remained with the Indians, and became connected with the tribe. The Indians having recovered the bell, carried it slung to a pole, through the forest; and it now hangs in the church steeple at St. Regis.

Lake St. Francis

is the name given to the St. Lawrence for a distance of 40 miles, between Cornwall and Coteau du Lac, where it widens considerably, and is interspersed with a large number of islands. At Coteau du Lac the river grows narrower again; and the Coteau Rapids (two miles long), the Cedars (three miles), the Split Rock, and Cascade Rapids are passed, the river descending $82\frac{1}{2}$ feet in 11 miles. There is a canal 11 miles long around these rapids, at the lower end of which is the village of Beauharnois. In the expedition of Gen Amherst, a detachment of 300 men, that were sent to attack Montreal, were lost in the rapids near this place. The passage through these rapids is very exciting. There is a peculiar motion of the vessel, which in descending seems like settling down, as she glides from one ledge to another. In passing the rapids of the Split Rock, a person unacquainted with the navigation of these rapids will almost involuntarily hold his breath until this ledge of rocks, which is distinctly seen from the deck of the steamer, is passed. Near Beauharnois, on the north bank, a branch of the Ottawa enters into the St. Lawrence. The river again widens into a lake called St. Louis. From this place a view is had of Montreal Mountain, nearly thirty miles distant. In this lake is Nun's Island, which is beautifully cultivated, and belongs to the Grey Nunnery at Montreal. There are many islands in the vicinity of Montreal belonging to the different nunneries, and from which they derive large revenues. At Lachine, nine miles above Montreal, the celebrated Lachine Rapids, short, but the roughest and most dangerous on the river, begin. The descent is 44½ feet in eight miles. Here the passengers crowd forward, and peer anxiously ahead and on every side, for the first glimpse of the long-expected, half-feared rapids. Just at the head of these rapids, a little Indian village, Caughnawaga, is seen on the right bank of the river. Here steam is shut off, and the boat comes nearly to a stand-still. A birch canoe puts out from the shore, with two men in it. It comes alongside; and a brawny, dark-skinned old man, in a picturesque garb, comes aboard. It is old Baptiste, the Indian pilot, who has for over 40 years piloted steamers through these rapids. He takes his place at the wheel, rings the bell to go ahead, and, aided by four or five powerful men, he steers the boat through the foaming, boiling surges, and past the ugly ledges that threaten to wreck her. The rapids safely passed, we shoot under the Victoria Bridge, and are soon moored to the magnificent pier at Montreal. These extensive piers, quays and walls of gray limestone, which border the entire river front, are among the finest in the world, and we gain a fair view of them as we land from the steamer. Passengers for Quebec and other points down the river are transferred direct to one of the fine steamers about starting, and then our steamer is warped into the locks at the foot of the Lachine canal, raised to the upper level, admitted to the basin, and we land on the broad quay, where a host of cabs, omnibuses and other vehicles are in waiting to convey us to our hotel.

CHAPTER XI.

Montreal and Its Environs.

MONTREAL, the Queen city of Canada, is built upon the island of the same name, formed by the confluence of the two mouths of the Ottawa river with the St. Lawrence, which is here a mile and a half wide. The island of Montreal contains 197 square miles, and from its fertility and beauty is called "the garden of Canada." The city takes its name from the mountain which towers behind it, and which Jacques Cartier, in 1535, named Mount Royal. At that time the site of the city was occupied by a walled Indian Village called Hochelaga. In 1603, Champlain brought hither a small colony of Frenchmen who settled here. In 1642, M. de Maissonneuve and his associates having bought the island, and dedicated it to the Holy Family, landed here and named the city, "Ville Marie de Montreal." It has now a population of about 150.000, is one of the principal commercial cities of the Dominion, and by far the most attractive on the scores of natural beauty and elegance of its buildings. The streets are straight, intersect generally at right angles, and present all the characteristics of streets in the great American cities. Along McGill, Great

St. James, and Notre Dame streets, the principal retail business thoroughfares, are many fine stores which make attractive displays of goods, while on Dorchester, St. Catherine and Sherbrooke streets, (the latter the Beacon street or Fifth avenue of Montreal), are numbers of princely residences. No city affords more comfortable, handy and cheap facilities for seeing the sights than Montreal. Hundreds of light, commodious and attractive one-horse hacks, which convey four persons with ease and speed, stand on the streets waiting for fares, and the prices charged by the hour or the trip are very reasonable. If a driver, who can talk English and who is disposed to point out places of interest, be secured, which can easily be done, a drive about the city, for half a day, will be found most delightful as well as instructive. But before minutely considering the objects of interest, we shall desire to find out a good hotel for our stopping place, and we cannot do better than to take the omnibus of the Ottawa Hotel, or to call a cab and tell the driver to land us there. This house having been enlarged and improved, will now accommodate over 350 guests. The Ottawa Hotel covers the entire space of ground between St. James and Notre Dame streets, and has two beautiful fronts. The house has been thoroughly refitted, and furnished with every regard to comfort and luxury; has hot and cold water with baths and closets on each floor. The aim has been to make this the most unexceptionable first-class hotel in Montreal. Messrs. Browne & Perley, the proprietors, have had long experience in first-class hotels in the United States and Canada; and guests can be sure of every attention and comfort. The St. Lawrence Hall, corner of St. James and St. Francois Xavier streets, is a large hotel much affected by English tourists, and the St. James, on the street of the same name, fronting Victoria square, is a quiet and comfortable house. But the Ottawa will be most satisfactory to Americans, being kept in the style of hotels in "the States," and provided with all modern conveniences.

Public Squares and Buildings.

Victoria square, in the center of the city, at the intersection of McGill and St. James streets, is a pretty enclosure with a fountain in the center, and fronting St. James street, stands a bronze statue of Her Majesty. This admirable work of art, was erected on the 21st of November, 1872, and presented to the City by His Excellency the Governor General. The cost of the statue, including that of the pedestal, — the gift of the Corporation, — was $13,000. The Place d'Armes, a pretty little garden enclosed with an iron fence, fills a square between St. James and Notre Dame streets, and upon it, (on the latter street), fronts the magnificent cathedral of Notre Dame. The Champ de Mars, on Craig street, at the end of St. James, is a famous promenade for citizens and strangers, being the general parade and review ground of the military, and is frequently enlivened during the summer evenings by music from the fine bands of the regiments. Viger Square near the Champ de Mars, is beautifully laid out into a garden, with conservatory, fountains. The Place Jacques Cartier, a broad, steep street, running from Notre Dame to the Bonsecour pier, is surmounted at its highest point by a tall column known as the Nelson monument, which was originally quite pretentious, but is now in a rather dilapidated condition. Among the public buildings worthy of especial note, are the new Court House, on Notre Dame Street, and directly opposite to Nelson's Monument, of elegant cut stone, in the Grecian-Ionic style; the Post-Office, on St. James Street, a beautiful cut-stone building; the Merchants' Exchange, on St. Sacrament Street; the Mechanics' Institute, a very fine building, situated on St. James Street, of cut stone, three stories high, built in the Italian style; the Mercantile Library Association Building, Bonaventure Street; the Bank of Montreal, Place d'Armes, St. James Street, opposite the Cathedral, an elegant cut-stone building of the Corinthian order; the City Bank, next to the above, in the Grecian style of cut stone; the bank of British North America, St. James Street, next to the Post Office, a handsome building of cut stone, in the Composite style of architecture; Molson's Bank, St. James Street, a handsome structure; the Bonsecours Market, on St. Paul and Water Streets, a magnificent edifice in the Grecian-Doric style, cost about $300,000, and contains the various offices of the city; the McGill College, an institution of very high repute, founded by the Hon. James McGill, who bequeathed a valuable estate and £10,000 for its endowment; the Old Government House, Notre Dame Street, now occupied as the Normal School; the Barracks; the Custom House, St. Paul Street; Hotel-Dieu Hospital, Sherbrooke Street; and many others. The Lachine Canal is among the public works of which the city may feel proud.

Churches and Religious Institutions.

Of these there are many and notable. The French Cathedral of Notre Dame, fronting on the Place d' Armes is perhaps the one which first attracts the visitor. It is the largest on this continent, seating 10,000 persons. It is 255 feet long by 144 wide and has two towers on its front, each 220 feet high. From the top of one of these towers, to ascend which a fee of 25 cents entitles the visitor, a magnificent view is gained of the city, the river, spanned by the Victoria bridge, and alive with shipping, the islands and the American shore. In this tower hangs the bell "Gros Bourdon," the largest in America, weighing 15 tons, and in the other tower is a fine chime of bells. The Church of the Gesu, or Jesuit church, on Bleury street, is famed for the magnificence of its interior decorations, especially its frescoes, portraits of saints and altar-pieces. The nave is 75 feet high and the roof is sustained by rich composite columns. St. Patrick's church on Lagauchetiere street, is 240 by 90 feet, with a spire 225 feet high. Christ Church Cathedral (English) a splendid Gothic church on Catherine street, is cruciform, with a stone spire 224 feet high from the centre of the cross. It is of Caen and Montreal stone, and is lighted by stained glass windows, several of which are very beautiful memorial offerings. The roof is sustained by elegantly carved Caen stone pillars. In the vestry is a bust of Bishop Fulford and a painting of the Rev. John Bethune, for many years rector. In the enclosure outside, is a fine monument to Bishop Fulford. Adjoining is a chapter-house and library. There are besides almost innumerable churches — Episcopal, Catholic and Presbyterian predominating, all of stone and all costly and handsome. On Dorchester street, an immense new Catholic cathedral, to be a copy of St. Peter's at Rome, but smaller, is slowly building, and near by is the Bishop's Palace. Of the nunneries, that of the Grey Nuns on Guy street, near Dorchester, is most visited. It is customary to be here a few minutes before noon and take seats in the chapel, where, at the stroke of 12 the nuns enter in procession, kneel and chant the prescribed prayers, in a subdued, sing-song tone of voice that is unspeakably depressing. Afterwards, visitors are conducted over the immense building, along acres of halls with floors scoured to the whiteness of snow, into the departments of the paupers, imbeciles and foundlings, through the artificial flower rooms, the laundry and other departments.

The Drive Around the Mountain.

By far the most delightful excursion is that around Mount Royal, over a splendid macadamized road, making a trip of nine miles. Ascending on the east side and passing around to the north and west, a magnificent view is gained of the valley of the Ottawa, the hamlet of Charleroi with its convent and church, with other villages nestling in the fertile valley, and the beautiful villas here and there dotting the mountain side. On the back side of the mountain are several tanneries, around which quite a little French hamlet has sprung up, and near by is the Half-Way House, where your driver will not, probably, refuse to take a glass of "'alf and 'alf," or "shandy gaff" at your expense. [N. B. — If you ever "take anything," yourself, you will find it good.] On the way towards town on the west side, the entrance to Mount Royal Cemetery, the French burying ground, is passed, and it is often visited by strangers. The sight of a procession of hearses, which have come here with as many funerals, racing on their way back to town, seems odd to any one but a native, yet it is often seen. The hearses for children are here very handsome — snow white with figures of angels at the corners and profusely ornamented. A little further down, the carriage stops and you step out upon a broad platform, whence the outlook upon the broad St. Lawrence, with Nun's Island rising far away in the distance, is very fine. To the south, lie the blue hills of Vermont, and at our very feet nestle the imposing buildings of Montreal. Returning to the city, it will be worth while to drive through Sherbrooke street and admire them any beautiful residences with their extensive grounds. Those of Mr. Brydges, Mr. Redpath and Sir Hugh Allan, (the last far up the mountain side and reached by a long private avenue), are among the finest in this vicinity. Still further down town, Dorchester street has many splendid mansions, — that of Mr. Harrison Stevens, the owner of the Ottawa Hotel property, being perhaps the finest. The house is completely surrounded by a lovely park, with dense folliage, shady walks, clumps and mounds of flowers, statuary, fountains, etc. On the corner of University street is the St. James Club, a large and elegant establishment. In the winter, when the mountain roads are deeply covered with snow, and when the fences are invisible beneath the icy crust, a favorite sport is "snow-shoeing." Clubs of young men with their snow shoes start out of a moonlight evening for a tramp of a dozen or fifteen miles, and return with tired limbs, flushed cheeks and prodigious appetites. Coasting down the mountain on sleds is also a favorite Montreal amusement, as is skating.

Next in interest to the drive around the mountain is that on the Lachine road, leading to the village of that name, nine miles from the city. The road is directly along the banks of the river, presenting scenery of unsurpassed beauty and grandeur. It is a lovely drive. If the proper hour is selected, a view may be had of the descent of the steamer over the rapids. Another favorite drive in the immediate vicinity is to Longue Pointe, being in an opposite direction from the last, and down the bank of the river.

The River Commerce of Montreal.

An immense volume of wealth pours into Montreal, through the St. Lawrence, that great funnel with its mouth to the northeast, and its outlet in Lake Ontario. Besides the lines of steamers above Montreal, there are the Richelieu Company's steamers "Quebec" and "Montreal," for Quebec, daily; the Union Steamboat Company's steamers "Abyssinian" and "Athenian" for the same point; the Ottawa River Navigation Company's boats "Peerless," "Prince of Wales," "Queen Victoria," and "Princess," for Ottawa, twice daily; the Allan Line Ocean Steamships for Quebec, Liverpool and Glasgow, twice each week; the Dominion Line Steamships for Quebec and Liverpool, weekly, and the Temperley Line of Steamships for Quebec and London, every two weeks. The Grand Trunk Railway radiating in every direction except north, affords land communication with the western cities, Boston, New York and Quebec.

Down the River to Quebec.

By far the most delightful voyage from Montreal is that down the St. Lawrence to Quebec, and if we wish a good night's rest, and fine views of the river at either end of it, (the night's rest, not the river,) and a sight of Quebec, by sunrise, we will embark, in the afternoon, on either of the Richelieu or Union steamers named above, and continue down the river, the first place passed being Longueil, a small village on the south bank of the river, 3 miles below Montreal. Fifteen miles below Montreal is Varennes, situated between the St. Lawrence and Richelieu Rivers. It is connected with Montreal by a line of steamers, and is attracting attention on account of its mineral springs. The first stopping place is at Sorel, forty-five miles below Montreal, at the mouth of the Richelieu, having in the vicinity good fishing, and snipe-shooting. Just below, the river expands into Lake St. Peter, 25 miles long, and 9 miles wide. Half way between Montreal and Quebec is the town of Trois Rivieres, at the mouth of St. Maurice. This is one of the oldest towns in Canada, is the residence of a Catholic bishop, and has a cathedral and convent. Thirty miles from Trois Rivieres is the mouth of the Shawenegan River; and a little above, on the St. Maurice River, are the Shawenegan Falls, where the water leaps down 150 feet perpendicularly. The last place at which steamers stop before reaching Quebec is Batiscan. In passing down the St. Lawrence from Montreal, the country upon its banks presents a sameness in its general scenery, until we approach the vicinity of Quebec. The villages and hamlets are decidedly French in character, and are generally made up of small buildings, the better class of which are painted white, or whitewashed, with red roofs. Prominent in the distance appear the tin-covered spires of the Catholic churches, which form the central figures of each of the villages. As we near Quebec, we see the banks of the north shore of the river become more bold and finally precipitous, and the houses nestle at the foot of the bluffs at the edge. The rafts of timber afford a highly-interesting feature on the river. On each a shed is built for the raftsmen, some of whom rig out their huge, unwieldy craft with gay streamers, which flutter from the tops of poles. Thus, when several of these rafts are grappled together, forming, as it were, a floating island of timber, half a mile wide and a mile long, the sight is extremely picturesque. Myriads of these rafts may be seen lying in the coves at Quebec, ready to be shipped to the different parts of the world. In the early morning, we look out upon a wall of rock rising above us on the left bank, like a mountain range, and at its base acres of these rafts are moored. Soon we round a lofty bluff to the left; we catch a glimpse of a mighty cliff, capped by a noble fortress, and in a moment Quebec is before us. In five minutes more we land on the quay.

The City of Quebec and its History.

Quebec, the great historic fortress of North America, what associations of romance and tradition throng the mind of the visitor who first gazes upon "the walled city!" Even in that name, "walled city," there is something so mediæval, so old-worldish, that it seems more a dream than a reality that it is before us. The first view of the historic city, in the radiance of early morning, is most inspiring and brilliant. The bold, massive headland of Cape Diamond juts a wall of eternal rock into the river. Perched on its summit is the gray old citadel, frowning down upon the river and the town, which latter clings to the sides of the mountain and clusters round its base, as if ever vainly trying to creep to the top, and ever slipping down. The roofs and steeples, sheathed in glittering tin, shine in the sun as if of burnished gold, while the green slopes of the glacis leading up to the citadel look like a velvet curtain ready to be drawn over the dazzling show. With the landing at the quay all the romance vanishes. You awake to the fact of narrow, dirty streets, importunate cabmen and shaky calashes, which make one seasick to look at, and nearly shake one into an omelet to ride in. By a street only less steep than the roofs of the houses, you are conveyed part way up the cliff, to the "upper town," which hangs midway between the citadel and the river, and which only the most persistent belaboring and a shocking waste of artistic profanity (in French Canadian *patois*) can induce the horses to surmount. It is evident there is no S. for the P. of C. to A. in Quebec. You reach your hotel, the St. Louis, on the street of the same name, or the Russell House on Garden street, both kept by Willis Russell, Esq., and excellent houses; and after a hearty breakfast, are ready for a tour of the city or a drive into the environs. You will have plenty of offers of calashes, coaches and cabs, but the best plan is to select a nice looking,

English-speaking driver, with a clean one-horse hack, and make a solemn compact with him to drive you wherever you want to go and be as long as you desire about it, for so much an hour. You can make the terms easy and can afford a long drive.

A Short Chapter of History.

The city of Quebec was founded by Samuel de Champláin, in 1608. In 1622 the population was reduced to fifty souls. In June, 1759, the English army under General Wolfe landed upon the Island of Orleans. On the 12th September took place the celebrated battle of the Plains of Abraham, which resulted in the death of Wolfe and Montcalm, and defeat of the French army. A force of five thousand English troops under Gen. Murray was left to garrison the fort, and in April following was besieged by the Chevalier de Levis and his re-organized French army. In a sortie of Murray, he lost 1000 men and 20 cannon, and had to retire again within the walls. The coming of an English fleet raised the siege, and the Treaty of Paris gave Quebec to England. On New Year's Eve, December 31, 1775, Generals Montgomery and Arnold, with a force of New England and Continential troops, attempted to storm Quebec, but Montgomery fell at the head of the forlorn hope, (a sign on the rock above points out the spot) and Arnold's men were hemmed in and a part of them captured. Since then the city has dwelt in peace, though its magnificent fortifications have been preserved and stregthened until within a few years, the home government has withdrawn the regular troops, and the wall has been partially dismantled.

The Walls and Fortifications.

Quebec is nearly triangular in form, built upon the crest and around the base of Cape Diamond, a lofty headland rising from the intersection of the St. Charles and St. Lawrence rivers, in an almost perpendicular cliff from 100 to 200 feet high. On the brow of this cliff the walls are built of solid hewn stone, bastioned and loopholed, and at the angles and salient points, batteries of artillery are placed. Two sides of the triangle of Cape Diamond are thus guarded. On the landward side, far up above the city proper, the triangle is completed by a line of wonderfully strong works, consisting of ramparts, ditches and outworks, while at the corner, nearest the St. Lawrence and on the very apex of Cape Diamond, 350 feet above the river, stands the Citadel. Since the regulars were withdrawn, a small body of Dominion artillery garrisons the Citadel, and visitors are shown over it. From the northerly bastion the view of the river, the new fortifications at Pointe Levis on the south shore; of the city and suburbs and the valley of the St. Charles, is magnificent. The works of the citadel are the wonder of engineers and military men. Underneath the ramparts on which we stand are casemates, now used as barracks, and beneath them yet another tier; at each angle is an Armstrong hundred pounder, and all around heavy ordnance frowns through dark embrasure. The provisions for the storage of ammunition and supplies are wonderful, and the number of men which can here be accommodated and utilized as a garrison seems incredible to any but military visitors, Outside the Citadel are the "old French works," now in ruins, and a line of martello towers, four in number. These extend across the peninsula, and are connected by underground passages with the Citadel. Originally, access from any direction to the "Upper Town," as the enclosure inside the walls is called, was through five gates,—massive stone arches, with iron doors, protected by powerful works and armaments. Of these, the St. Louis gate, near the covered way communicating with the Citadel, was the entrance from the Plains of Abraham, by the Grande Allée; the St. John's Gate, on the street of the same name, towards the St. Charles, was the entrance way of the St. Foy road; the Palace Gate, on the street of the same name, led down into the Lower Town near the St. Charles; the Hope Gate, on St. Famille street, 900 feet east of the Palace Gate, led to the wharves of the harbor, and the Prescott Gate, on Mountain street, barred the way up from the Lower Town market and the steamboat piers. All these gates, except the St. John's, which was rebuilt in 1867, have been removed on account of their obstruction of business. The walls have also been lowered, and a portion of the armament removed.

The Public Buildings of Quebec.

The Parliament House, where sits the Provincial Legislature, is a large old building, overlooking the former site of the Prescott Gate. On the ramparts in front of this building is the Grand Battery of (22) 32-pounders. But a short distance south, is the Durham Terrace, a broad, level platform of wood, with an iron railing on the river side, which rests on massive cut stone, on the very verge of the cliff. These foundations were those of the Chateau St. Louis, the first building in Quebec, of which Champlain laid the corner stone May 6, 1624. January 23, 1834, the castle was burned, and by Lord Durham's order the ruins were cleared away and this terrace built. It is the fashionable promenade, and in the evening the views of the river are very fine. From the terrace we may look down into the houses in the Lower Town, and might almost toss a stone into their back windows. The Governor's Garden is near by. It is a rather neglected little park, with gravel walks and a few benches. Its principal feature is a plain, tall obelisk, known as the "Wolfe and Montcalm's Monument," the foundation stone of which was laid by Lord Dalhousie, with imposing ceremonies, on Thursday, November 15, 1827. The monument is from a design by Major Young of the Seventy-Ninth, and cost upwards of £700. Being 65 feet in height, it is a striking object from the river, rising as it does clear from the garden. It bears two inscriptions; one of them by Dr. J. Charlton Fisher, as follows: —

"Mortem Virtus Communem
Famam Historia,
Monumentum posteritas
Dedit."

The Lower Governor's Garden is separated from the one just described by Rue des Carrieres, and has a masked battery for its principal feature. The Place d'Armes is a neat little garden, with a fountain, near the Durham Terrace, and the Anglican Cathedral, a large, plain old structure of stone, in which are the tomb of the Duke of Richmond, who died while Governor General of Canada, in 1819, and the monument of Bishop Mountain. The church has a fine communion service presented by George III, and a chime of bells, which the "artist" in charge jangles most horribly on Sundays. It contains many memorial tablets, mostly to British officers who have died while on duty here. The Catholic Cathedral of the Immaculate Conception fronts the Market Square near by, and is a fine old structure with some valuable pictures, among which one of Christ by Van Dyck is the finest. The Seminary, founded by Bishop Laval, and the University which bears his name, join the Cathedral on the northeast. The University has a library of 50,000 volumes and a splendid museum, and the Seminary has a quaint old chapel with some fine paintings. There are about 400 students, who wear a peculiar uniform. The market square is worth visiting on market days, for the quaint, old country pictures it presents. On the other side of the square are the Jesuit Barracks, so called, originally built for a college in 1646, but since 1759 used as quarters for troops. On Garden street, near by, is the Ursuline Convent, which with its gardens covers seven acres. In the chapel rest the remains of Montcalm.

The Suburbs of Quebec.

Passing out through the St. John Gate, we traverse the suburb of St. John, much the most modern and thriving looking part of the city. It lies on the high ground outside the walls, and contains many fine dwellings, stores and modern churches. Steep streets lead down to the suburb of St. Rochs, where are the Marine and General Hospitals, imposing and extensive institutions. This part of the city was ravaged by fire and almost annihilated some years ago.

By the Grande Allée, we traverse the historic Plains of Abraham, a lofty plateau on the landward slope of Cape Diamond, outside the citadel. At the time of the great battle identified with the name, the whole heights, or plains as they are indifferently called, extended from the walls to the woods of Sillery and St. Foy, and were bounded on one side by the St. Lawrence and on the other by the St. Charles. Since then, great encroachments have been made; the suburbs of St. Louis and St. John occupy great portions; and the name Plains has for a number of years been confined to an enclosed place, which is now a race-course. *Sic transit!* Near by is the jail, a substantial, cosey looking structure, and near that, the Wolfe monument—a plain circular column, rising from a square pedestal, and surmounted by a sword and helmet. On the one side of the pedestal is an inscription, as follows:

HERE DIED
WOLFE
VICTORIOUS
Sept. 13
1759.

And on the other side

THIS PILLAR
WAS ERECTED BY THE
BRITISH ARMY
IN CANADA, 1849.
HIS EXCELLENCY
LIEUTENANT GENERAL
SIR BENJAMIN D'URBAIN
G. C. B., K. C. H., K. C. T. S., ETC.
COMMANDER OF THE FORCES.
TO REPLACE THAT ERECTED BY
GOVERNOR GENERAL LORD AYLMER, G. C. B.,
IN 1832,
WHICH WAS BROKEN AND DEFACED,
AND IS DEPOSITED BENEATH.

The Cap Rouge road, on which we now enter, is a pleasant drive, bordered by fine villas. Spencer Wood, a magnificent park, with concrete driveways half a mile long, lighted all the way by street lamps, and with all the pomp of a feudal domain about it, is the residence of the Provincial Lieutenant Governor. Knowing this, it is a disappointment to find that his mansion is an old, yellow barrack, with no pretensions to beauty, outwardly, at least. Visitors often cross from this point to St. Foy road and return to the city by the St. John's Gate, passing on the left the "Monument aux Braves," a tall column, surmounted by a statue of Bellona, presented by Prince Napoleon. This monument marks the scene of the second battle of the Plains, where Murray was defeated by De Levis as already described, and was erected in 1854 by the French residents, over the remains of hundreds of their blood who fell in that conflict.

The Indian village of Lorette, near the falls of the same name, nine miles inland, is often visited; so also are the Chaudiere Falls, 350 feet wide and 150 feet high, 18 miles from Pointe Levis, on the south side of the river.

The Falls of Montmorenci.

But the most interesting of the wonders near Quebec, are the Falls of Montmorenci, eight miles from the city by the Beauport road. Crossing the St. Charles, we follow the bank of the St. Lawrence at some distance, passing several handsome villas and chateux, and an old mansion house, now in ruins, where Montcalm had his head-quarters at one time. Beauport is a long village, on both sides the road, with no particular beginning nor ending, and notable for its Frenchness in cottages, people and dialect; also for the crowds of children who run beside your carriage, with bunches of flowers which they expect you to accept — in return for small siver coin. The Falls are seen by dismounting and traversing a field a short distance to a pavilion, which brings us face to face with the cataract. Here the Montmorenci river plunges down a perpendicular precipice 250 feet high, into a tremendous yawning gorge, with a roar that is heard for a long distance. So abrupt is the fall that the water is beaten into perfect foam, and looks more like a sheet of wool hung on the face of the rock than a cataract of water. Just above the Falls are the towers of a suspension bridge, which spanned the river, but soon fell and pitched three persons into the chasm below. They were killed.

Ste. Anne, 24 miles below Quebec, is visited by steamer if one chooses to make the excursion. Its objects of interest are several picturesque falls, lofty mountains and a pretty little pilgrimage-church, where the relics of Ste. Anne are exhibited, and whereby many wonderful cures of the sick are reported. There are also many other excursions from Quebec, if we have time and inclination. They will be suggested by the driver.

The Lower St. Lawrence and the Saguenay.

The tour of the Lower St. Lawrence and the Saguenay is made from Quebec by the steamers of the Saguenay Company, whose office is on St. Louis street, opposite the hotel, Stevenson and Leve, agents. The clerk in charge, Mr. Stocking, is an American and a white man, which is praise enough. The boats leave every morning except Sunday and Monday, and make the round trip in 48 hours. Or we may take the Grand Trunk to the mouth of the river, and then go up by steamer to Grand Bay (or Ha, Ha! Bay, as it is also called.) The Saguenay river is formed by a junction of two outlets of the St. John Lake, a body of water covering 500 square miles, and lying in the wilderness 125 miles northwest of Tadousac. Up towards the lake there are magnificent cascades, where the water dashes along between banks of solid rock from 100 to 1,000 feet high. Ha-ha Bay, which is 60 miles from its mouth, affords the first landing and anchorage for steamers. The Saguenay boats, Union, St. Lawrence and Saguenay, are small, but comfortable and well found boats, and the fare is excellent. From Quebec to the mouth of the Saguenay, the scenery of the lower St. Lawrence is magnificent. On a late summer's day, when the sun shines brightly, owing to some peculiarly atmospheric condition, the landscape is overspread with a soft haze, and the view across the glassy water, to the rolling hills and gentle slopes of the southern bank strikingly resembles, in its mellow, dreamy softness, an Italian landscape. On the north side of the river, which the boat hugs most of the way, the scenery is grand and rugged, a lofty, precipitous mountain range extending all along; the cliffs, rising abruptly from the water's edge, being generally densely wooded with evergreens, and frequently threaded with sparkling waterfalls. At St. Paul's Bay, we make a stop alongside a fishing schooner, anchored in deep water — the shoals inshore prohibiting a nearear approach — exchange mails, land a passenger or two, and are off again. At Eboulements and Murray Bay, stops are also made. Then we dart across the river to Riviere du Loup, on the south shore, where is a popular Canadian watering-place, and which also serves as the landing for Cacouna, the Newport of the Dominion. Here we meet the other boat of the line, on her return trip from the Saguenay. Again, crossing the river abruptly, we are soon alongside a little pier at Tadousac, just within the mouth of the Saguenay, and the captain gives us an hour's "leave on shore." If Cacouna be the Newport, Tadousac is the Long Branch of the Dominion. Here Lord Dufferin, the Governor-General, has a summer home, emulating the Chief Magistrate of the States in the choice of a summer capital, by the sounding sea. The St. Lawrence, at this point, is quite a sea in size, and, as its waters are salt, and the tides flow and ebb, "sea bathing" is one of the advertised attractions. There is a fine hotel at the head of a crescent-shaped bay on the St. Lawrence side of a jutting point, and quite a group of Summer cottages on the bluff above.

The Mysterious River.

But we have come to see the Saguenay and not to revel in the enjoyments of Tadousac, and soon we are speeding up that mysterious river, just as the sun is sinking behind the

SAGUENAY RIVER, TADOUSAC.

grand and gloomy cliffs which confine it. We have noticed, even before our landing at Tadousac, that the water has grown as black as ink, almost, and we find it to rival Day & Martin, as soon as we are fairly inside the point. This peculiarity, they tell us, is due to the river's flowing through hemlock and tamarack swamps. We observe that the churning of our paddle wheels produces, not white froth, but something resembling coffee cream. We have been prepared for a wild and startling panorama, for fathomless waters and precipitous, overhanging walls of rock, but the reality surpasses the description, and, when the moon comes out and throws her silver radiance across the gloomy depths beneath us, and tinges with pale splendor the bare, gray cliffs above, the effect is strangely weird and unreal. The banks present a continual succession of pictured rocks and towering precipices, "It is as if the mountain range had been cleft asunder, leaving a horrid gulf 60 miles long and 4000 feet deep through the gray mica schist." Among the points of greatest note on the Saguenay, may be named Statue Point, an immense perpendicular rock below Ha-ha Bay, which rises 600 feet, with sides as smooth as if polished by a skillful workman. "Statue Point has a huge Gothic arch, opening into a vast cave, which it is said, the foot of man never trod. Before the entrance to this black aperture a gigantic rock, like the statue of some dead Titan, once stood. A few years ago, during the winter, it gave way; and the monstrous figure came crushing down through the ice of the Saguenay, and left bare to view the entrance to the cavern it had guarded perhaps for ages." Beyond this is the vast Tableau Rock, a sheet of dark-colored limestone, some 600 feet high by 300 wide, as straight and almost as smooth as a mirror.

Ha, Ha! Bay

About midnight we arrive at Ha, Ha! Bay and take on wood, which generally occupies the rest of the night, as the French deck hands are not used to hurrying, and besides, the passengers want to see the bay by early morning light. Ha, Ha! Bay is a broad expanse of smooth water, shut in on all sides save one by rugged and almost inaccessible cliffs. This bay is a pocket, opening from the right bank of the river, and is said to be seven miles deep. At the bottom of the pocket is a strip of low land, which has been improved by the location of two little hamlets, named respectively St Alphonse and St. Alexis. The latter is the larger, but the landing is at the former. Each is a cluster of little cabins, in the centre of which is a small church. About a mile from the landing is a salmon river, where in the season the salmon may be seen by scores, leaping, like animated bars of silver, to the height of several feet, in the attempt to surmount the falls. This salmon river, like most in the Dominion, is rented out by Government, the lessee being a Mr. Price, who owns a vast lumber territory, up this way, and seems to number nearly all the inhabitants among his employés. About all the pay they get comes in the form of provisions and clothing from his stores.

To the Head of Navigation and Back Again.

Starting on our way again we proceed out of the bay and up the river, through a tortuous and difficult channel, to Chicoutimi, the head of navigation. Above Ha, Ha! Bay, the shores of the river are less abrupt, frequent fertile slopes and green valleys appear, and there are many habitations. Chicoutimi is the metropolis of this region. Here is Mr. Price's lumber *entrepot*, and here he has two of his stores. Large barques lie at anchor in the channel, loading with the staple commodity sawn at the mills on the Chicoutimi river, which here empties into the Saguenay. Chicoutimi consists of a long, straggling street, lined with small houses, mostly of logs, a barn-like frame church, a nunnery, the residence of the priest and those of Mr. Price and his brother; the last three being quite comfortable houses. It is a most lonesome and desolate metropolis, and its most lonesome and desolate feature is the little graveyard beside the church, with its wooden headboards, bearing French inscriptions in black paint. Upon the most sightly prominence overlooking the river, are rising the walls of a large stone building, which we are told is to be a Catholic college. Fancy a *college* in these wilds! After an hour or two spent in seeing the sights of Chicoutimi, we start on our return. Its incidents are not worth record till about 2 P. M., when we reach the wonder of the region, the great object which we have borne in mind during the whole trip, but which we have not before seen, as we passed it in the night on our up trip — Eternity Bay.

The Wonder of Wonders.

As we near this spot, the overhanging cliffs grow higher, the whole panorama wilder, and by degrees our eyes are educated to a proper appreciation of the great marvel. We are sailing in fathomless waters between walls of rock, towering hundreds of feet above us, and gradually increasing to near 2,000 feet from the water's edge, when just in the highest part the wall is cleft asunder, and a broad inlet makes into the land. That is Eternity Bay. The steam is shut off, the vessel's head turned to the right, and slowly she drifts into the bay. To our right is a Titanic cliff, at first descending in three gigantic steps to the water's edge, but, as we round the point, its face becomes one sheer, perpendicular surface of rock, veined and streaked with red and black, as if the solid mountain had been split asunder, and the very heart of the rock exposed. Its upper edge bears the shape of three domes, set side by side, justifying the title of Cape Trinity. Here we drift along, while the steamer's whistle is blown and a small cannon fired, the blasts and the report coming back to us in marvellously distinct echoes. A box of small stones is brought out, and we try to throw a pebble across the apparently insignificant expanse of water between us and the cliff, but in vain. Such is the deception in distance, due to the overpowering height of the cape, that the strongest arm fails to compass it. We could remain here for hours, and every moment gain a better appreciation of the majesty of the scene, but time fails, and the steam is again applied, and the boat's head turned riverward. As we pass out of the bay, Cape Eternity looms above us, the twin sentinel of the wonder we have just left, less rugged and harsh, but even more massive and higher by 100 feet. Its loftiest wall towers 1,800 feet perpendicularly above the dark water of the river, which is here said to be a mile and a quarter in depth. These are the marvels of our journey, and they passed, nothing remains to excite our wonder. The rest of our voyage is but a repetition of what we have already seen. We reach the mouth of the river at nightfall, and the next morning we again land at Quebec.

Other Trips from Quebec.

From Quebec, we may — if we desire a longer trip, and one which will give us a view of the wild and majestic scenery of the Gulf of St. Lawrence, and the boundless expanse of the North Atlantic — take passage by one of the splendid steamships of the Quebec and Gulf Ports Steamship Company, of which, also, Stevenson & Leve are agents, — the "Secret," "Miramichi," or "Georgia," — for Shediac, Pictou, or Charlottetown. On the trip we shall see Perce, with its famous arched rock rising from the waters, and affording a passage for the waves; the Isle of Orleans; Father Point; Chaleur, Miramichi, and Gaspé Bays. From Shediac or Pictou we may return to Boston by Intercolonial, European and North American, Maine Central and Eastern Railways, passing through St. John, N. B., Bangor, Augusta and Portland, Me. A more delightful winding up of a summer trip could not be imagined. Or we may cross the St. Lawrence to Pointe Levis, take the Grand Trunk Railway cars and, passing through Richmond Junction, continue our journey to Portland, Me., and thence home; or to Newport, Vt., and thence to the White Mountains or directly to Boston. Or we may return to Montreal and thence take one of the already described routes to Boston.

MAP OF THE WHITE MOUNTAIN REGION.

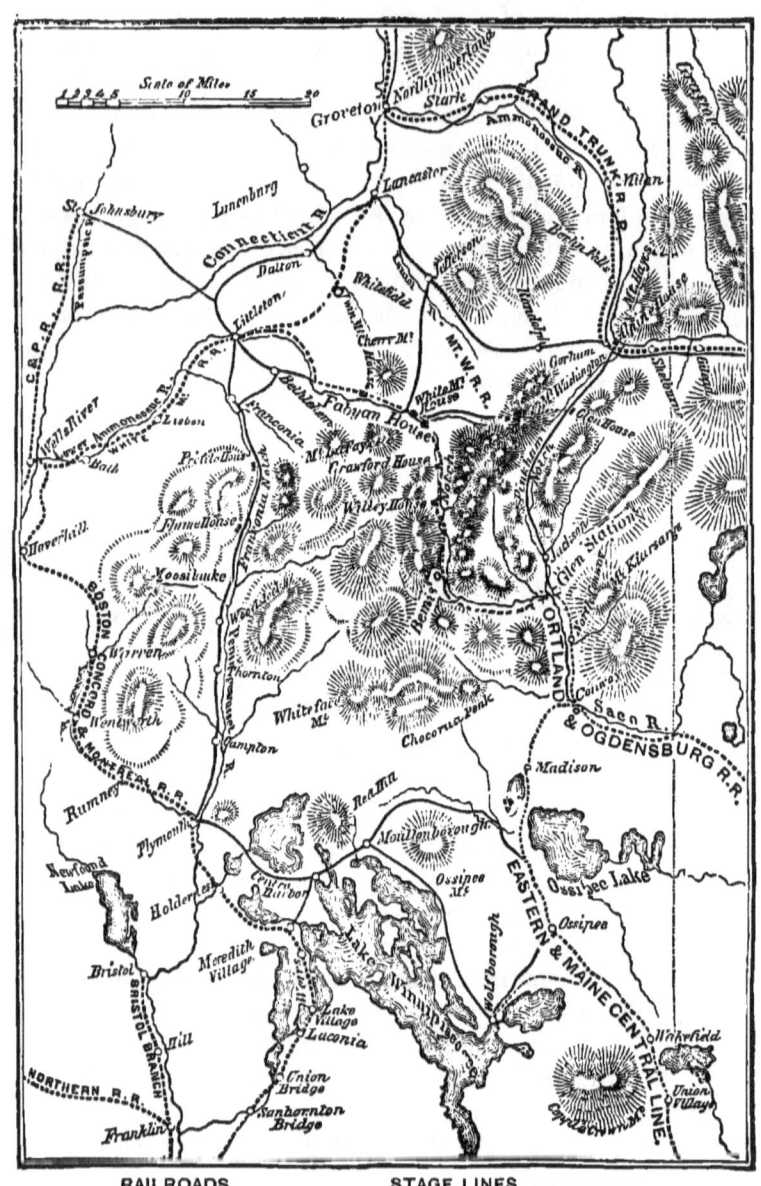

CHAPTER XII.

The White Mountains.

OF ROUTES to the White Mountains there are many. We have already in imagination traversed several; but we will now approach "the Switzerland of America" from the north, *en route* to our Boston homes. From Quebec, as we have already said, the trip is made via Newport, Vt., but unless one wishes to stay awhile at Lake Memphremagog, this route is not convenient, the connections being uncertain and far from "close." From Montreal, there are three routes from which to choose: one by the Grand Trunk Railway from St. Lambert *via* Richmond Junction to Littleton, 187, or Bethlehem, 199 miles from Boston; the second, *via* South-eastern and Connecticut and Passumpsic

Rivers Railroads from St. John's, down through Newport to Wells River, Vt., and thence up to Littleton or Bethlehem; the third by the main line, through St. Albans to Essex Junction, thence by Central Vermont down to White River Junction, then up through Wells River to Littleton or Bethlehem. Arrived at the railway terminus, — whichever of the two last-named stations we choose as our point of approach to the mountains, — we transfer ourselves to the six-horse stage-

THE PROFILE, OR "OLD MAN OF THE MOUNTAIN."

coach which conveys us to the Profile House, in the very heart of the Franconia Notch. This hotel stands on a plateau of level ground in the midst of the great hills, and at the very foot of Eagle Cliff, a towering crag, which seems to threaten the house below, and takes its name from the fact that a few years ago a pair of splendid eagles made it their home. The view down the Notch, with its sentinel peaks on either hand, is grand and imposing. Echo Lake is one of the noted features of the Franconia Notch, a diminutive but very deep pond, entirely environed by mountains. From its centre a voice, the notes of a horn, or the discharge of a fire-arm, will awake a perfect chorus of echoes many times repeated.

Profile or Cannon Mountain derives its names from the resemblance of a pile of rocks on its summit to a mounted cannon, 2,000 feet above the road, double that height above the sea level, and directly opposite Eagle Cliff, forming the western side of the Notch; and from the Profile on the southern extremity of its crest. This "Great Stone Face," immortalized in literature by Hawthorne, and familiar to all visitors as the "Old Man of the Mountain," is eighty feet from the point of the chin to the top of the forehead; and it is placed 1200 to 1500 feet above the road. Three masses of rock form this profile, which is clearly cut and entirely distinct, with a high, stern forehead, prominent nose, lips just parted, and a massive chin. At the foot of this mountain lies the lovely little Profile Lake, called also the "Old Man's Wash Bowl." Bald Mountain is ascended from the hotel by a carriage-road; and from its summit a fine view is obtained. Mount Lafayette is the giant of this range, towering 5,280 feet skyward, and pyramidal in form. Its ascent is long and tedious, by a devious bridle-path; but the view from its summit, taking in the whole mountain range and surrounding country, compensates for the fatigue. Walker's Falls, in the rear of the road, are a series of mountain cascades, leaping like a ribbon of silver down through a contracted gorge. The Basin is five miles south of the Notch, and lies near the roadside. It is formed by the action of the water of the Pemigewasset, which pours over a rocky ledge into a hollow in the solid granite. This hollow, by the incessant whirling of the water and the pebbles which it carries with it, has been worn into a perfect bowl, nearly circular, 45 feet in diameter, and 18 feet deep. The clearness of the water is such that the smallest objects on the bottom are clearly discerned, though its great depth gives it a bright green tint.

12*

THE FLUME, FRANCONIA NOTCH.

The Flume is perhaps the most famous, and is certainly not the least wonderful, of the curiosities in the Franconia Mountains. Imagine a solid mass of granite, split to the depth of fifty feet, and the perpendicular walls separated twenty feet, and you have an idea of the Flume. Through it pours a little brook; and a plank walk alongside enables the visitor to ascend its course several hundred feet. Near the upper end a huge boulder, which evidently lay on the surface when the rock was riven, has fallen into the chasm, until the sides, gently sloping inward, have caught and hold it suspended in mid-air. The Cascade, below the Flume, is a waterfall of more than 600 feet descent, gliding over the polished rock like a sheet of molten silver. The Pool is an enlarged edition of the Basin. It is about 150 feet wide; and the water is 40 feet deep. It is cut from the solid granite by the chisel of Nature. From the top of the rocky wall which surrounds it, its depth is about 150 feet. The Harvard Falls, also called the Georgianna Falls, are the most remarkable cascades in the vicinity. For nearly a mile they follow each other down the mountain; and the uppermost makes a flying leap of 150 feet sheer. Having thus "done" the wonders of the Franconia region, the tourist may follow the valley of the Pemigewasset down to Plymouth, and thence by rail to Lake Winnipiseogee, or may retrace his steps to Littleton or Bethlehem, and thence by rail to Twin-Mountain station, at the very door of the famous hotel of the same name, 203 miles from Boston.

The Twin Mountain House.

This first-class hotel, built in 1869-70, is pleasantly situated on a rise of ground on the bank of the Ammonoosuc River, commanding a fine view of the White and Franconia Mountains. To the right rises Mount Lafayette in all its grandeur; while to the left, and distinctly visible, is the White Mountain range; and towering above all is Mount Washington. Being centrally located, parties can visit many points of interest, and return the same day. Among these are the Crawford House, with its White Mountain Notch, Mount Willard, the Willey House, and numerous cascades, Mount Washington and its railway, Profile House, Littleton, Waumbek House, and Bethlehem. From this house it is but ten miles to the Crawford House (five by rail, five by stage); ten miles to the depot of the Mount Washington Railway, where cars are taken for a trip over the famed rail line to the crowning summit; 30 miles to the Glen House, and 28 miles to Gorham, by the Cherry Mountain road. Parties visiting the mountains should not fail of making the ascent of Mount Washington by its railway, which is a triumph of mechanical skill and engineering. Thousands of persons are annually carried up this road with perfect ease and safety. Comfortably seated in

TWIN MOUNTAIN HOUSE.

their cars, rising at the rate of one foot in three, new objects of interest come before the eye. Villages, rivers, lakes, and mountains continually burst upon the view until the summit is reached, when the beholder stands upon the highest point of land in this country east of the Mississippi. Parties desiring to descend the mountain on the east side, by the carriage road, will find carriages in readiness for the Glen House and Gorham. Coaches run to and from the Twin-Mountain House to all important points about the mountains, and to the Mount Washington Railway. Leaving the house at 7.30 in the morning, you reach the summit at 12 M.; returning, leave the summit at 3, P.M., and reach the house at 6.30, P.M. The Boston, Concord, and Montreal Railroad has built a spur-track to this house, with an extension to the Fabyan House. Passengers leaving Springfield, Boston, or Burlington, Vt., in the morning, arrive at 6.20, P.M.; and those leaving Wells River in the morning arrive here at noon. Passengers can take the cars at this house in the morning, and reach Boston, New York, Newport, Burlington, or Montreal the same day. For the pleasure of the guests, the proprietors have provided billiards, bowling, pleasure-boats, croquet grounds, and a good band during the season. Post and telegraph offices are located in the house; and "horses and carriages, with experienced drivers, are furnished for parties when desired," Those afflicted with "hay-fever," or "autumnal catarrh," will find comfort in the fact that the larger number of those afflicted with this disease, who have been here in years past, bear testimony to partial, and in most cases entire relief from this distressing malady. Messrs. A. T. and O. F. Barron are the proprietors. One feature of the management of the Messrs. Barron is worthy of note. A farm of twelve hundred acres near White River Junction, and the old homestead farm at Queechee, Vt., supply vegetables, milk, eggs and butter for their houses, and guests may be sure of always getting the freshest and best of farm and dairy products at their table.

WHITE MOUNTAIN RANGE, FROM JEFFERSON HILL.

The Crawford House.

The same firm are also proprietors of the famous Crawford House, at the head of the White Mountain Notch, the headquarters of the region. This splendid hotel, newly furnished, and provided with all the appliances of a modern resort, is so placed as to command from its spacious piazzas a grand vista down the wondrous White Mountain Notch, and views of the summit of Mount Willard, which is reached by carriage road from this point, and of the Elephant's Head, a singular mass of rock, projecting from the mountain side, and so perfectly formed that no one needs to be told what it is. A glistening seam of white rock simulates the tusk, while the massive head, pendulous trunk, and huge ears are represented by dark gray crags. From the Crawford House one may ascend Mount Washington by bridle-path, carriage-road, or railway, spend the night at the Mount Washington, Summit, or Tip Top House, and descend next day on the other side to the Glen House; or he may take the stage coach, via Cherry Mountain, over the Jefferson Hills, from which a wonderfully fine view of Mount Washington, from a new standpoint, is gained. Starr King's most enthusiastic descriptions were of this locality. From Jefferson, a ride of 20 miles around the base of Mount Madison brings us to the Glen House; or, if we prefer to make the journey from the Crawford House by another route, we take the stage-coach at the door, and are soon rattling down through the world-famous White Mountain Notch.

The White Mountain Notch.

This is a gorge, or rift, through the mountains, which affords a water course for the Saco river. On either hand the mountains tower to the height of 2,000 feet; and the carriage road is cut from the very mountain side, clinging as it were to the verge of the steep declivity, while far below the river brawls and babbles over its stony bed. In one place, called the " Gateway," the Notch is but 22 feet wide. An ex-

"GATE OF THE NOTCH."

tension of the Portland and Ogdensburgh Railroad, up through this Notch, opposite the carriage path, is in progress, and will probably be open to travel during this season (1875.) On the way down the Notch we pass the Flume, a narrow sluiceway, worn into the solid rock of the mountain's side, down which courses, with the swiftness of light, a mountain stream. A little farther on we see the Silver Cascade. This is one of the most charming waterfalls imaginable, and may be traced like a thread of silver winding down over the glassy rock from 800 feet above the road. Still farther down, between Mount

Webster and the Willey Mountain, we see at the right of the road the historic building, the Willey House.

Here, on the 28th of August, 1826, the Willey Family, nine in number, alarmed by the noise and sight of a terrific avalanche coming straight down the mountain-side towards the house, fled, but were overtaken and buried by the rushing mass. A huge rock back of the house divided the earth-slide,

PEABODY RIVER AND MOUNT WASHINGTON.

and saved the house. It has been greatly enlarged, and is now a place of entertainment. Sparkling Cascade and Sylvan Glade Cascade are pretty waterfalls below the Willey House.

Leaving the Saco Valley, below Sawyer's Rock, we turn to the east, and cross the Ellis River, getting a view of the Goodrich Falls, the most lofty and imposing cataract in the mountains. A mile farther on Jackson is reached, where are some very beautiful cascades on Wild-Cat Brook. The views of the mountains are very grand from this point. From Jackson we proceed nearly north, up the Ellis River, and through the Pinkham Notch, passing by the way the beautiful Glen Ellis Fall, where the water of the river pours down over a precipice 85 feet high, making a perfect arch of foamy spray; past the lovely Crystal Cascade, about the same height, and aptly described as an inverted plume; past the Emerald Pool, with its quiet beauty; Thompson's Falls, and the Garnet Pools, and soon arrive at the Glen House. This is one of the largest hotels in New England, having about 400 rooms, and is one of the most complete and luxurious in all departments.

From this point teams are provided for trips to the Crystal Cascade, Glen Ellis, Emerald and Garnet Pools, the Imp Mountain, Tuckerman's Ravine, — with its gloomy depth and masses of eternal snow, — West Branch, Mount Carter, and, grandest of all, the ascent of Mount Washington. The road, which was completed to the summit and opened for travel in 1861, is a smooth and well-built macadamized turnpike. The average grade is 12 feet in 100. There is no difficulty in the ascent, and no more discomfort than in the same amount of carriage-riding upon any of the mountain roads. The carriages are easy and comfortable, and have experienced drivers capable of giving information. These carriages are accompanied by baggage-wagons; and at the summit of the Mountain we may take the railroad down to the White Mountain Notch.

Climbing the Mountain.

Having passed through the forest that covers the base of the mountain, the road emerges on the mountain side near the "Ledge." Clay, Jefferson, Adams, and Madison are seen to the best advantage from here; and Starr King calls these mountains, seen from this point, "Nature's struggle against petrification, the earth's cry for air!" Rising from the Ledge, the road overlooks the valley of the Ellis and Peabody Rivers, and the Saco Valley, famed in song. Plateau after plateau does the road reach, each one, as we look up to it, seeming to be the last. Finally, after about three hours' ride, we reach the summit, 6,300 feet above the level of the sea.

"The first effect upon standing on the summit of Mount Washington is a bewildering of the senses at the extent and lawlessness of the spectacle. It is as though we were looking upon a chaos. The land is tossed into a tempest. But in a few moments we become accustomed to this, and begin to feel the joy of turning round and sweeping a horizon line that in parts is drawn outside New England. The diameter of this circle is 250 miles; and you are at the central point. As far as you can see, in every direction, are mountains holding up their faces to be kissed by the sun. There are lakes, rivers, villages, and roads no broader than a ribbon, stretching away so many miles that it makes one tired; but the warm blue mountains, chain upon chain, are over and above all. Upon these things do you look down; and you can look up — only at heaven."

North Conway, and its Beauties.

From the Glen House, a splendid drive down the Pinkham Notch and the Saco Valley brings us to North Conway, long famed as a summer resort and a favorite haunt of artists, — the most gifted pencils in the country having transferred its charming scenes to their canvas. North Conway lies just at the portal to the mountains, whose snow-capped peaks form the back-ground for the most delightful views. The Conway Intervale stretches away on either hand, a broad expanse of richest green, threaded by the sparkling Saco River. The beautiful village, nestled at the foot of grand old Kiarsarge, is a pleasing feature, with its neat white houses, well-kept roads, and general air of thrift. The numerous hotels and large boarding-houses are taxed to their utmost to accommodate their crowds of summer visitors from the cities. The principal hotel is the Kiarsarge House, kept for many seasons past by Thompson & Sons. This long famous house was greatly enlarged in 1872, and fitted up in unsurpassed style. The Intervale House, kept by Mudgett & Son, is located under the slope of Mount Pequaket, or Kiarsarge, is within a

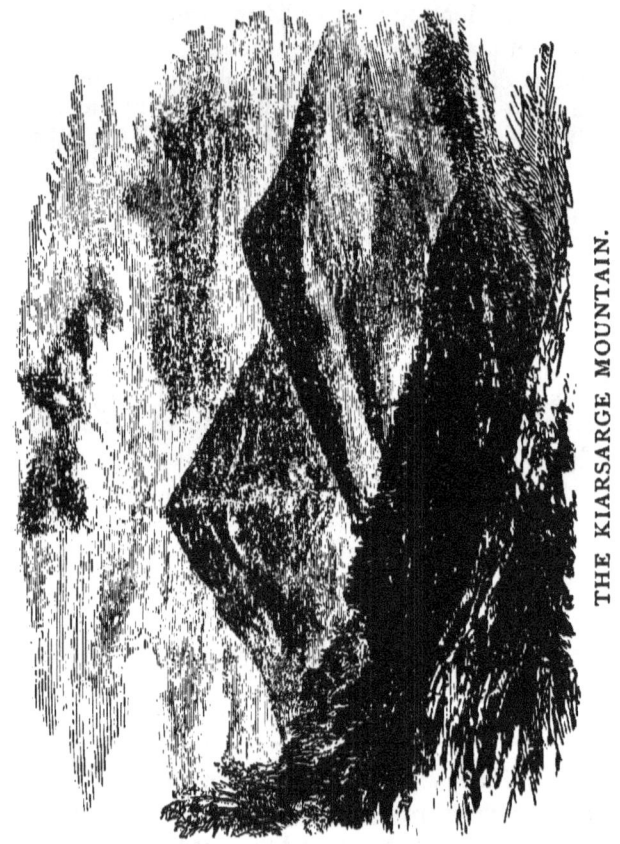

THE KIARSARGE MOUNTAIN.

beautiful enclosure of hills and surrounded by attractive points, easily reached by short walks from the hotel The buildings are comparatively new, and it will be found to be a centre of attractions for those who tarry at this point. It has about 100 sleeping rooms; has been newly painted throughout, and partly new furnished; has a nice hall for parties and hops, etc. The house is within a few rods of the Intervale Station of the Portland and Ogdensburg Railroad, to which the Eastern Road also runs its cars. The other hotels in the village are the North Conway House, M. B. Mason, proprietor; the Sunset Pavilion, kept by M. L. Mason; Mason's Hotel, F. H. Mason, proprietor; McMillan House, by J. McMillan; the Washington House, J. M. Gibson, proprietor; the Eastman House, kept by C. E. Evans. There are, besides, many boarding-houses for summer guests. The "Ledges," bold granite bluffs, nearly 1000 feet above the Saco, with the deep chasm known as the Cathedral; Diana's Bath, filled with limpid water, sparkling like crystal; Artists' Falls, in a shadowy glen of picturesque loveliness; Echo Lake, at the foot of Mote Mountain, and the "White Horse," — are a few of the objects of special interest in the vicinity of North Conway. The queenly Kiarsarge, a symmetrical paramid 2,367 feet high, is seen to the north-east, and to the west, Mote Mountain, with Chocorua's jagged peak in the distance; while the curves of Rattlesnake Ridge, and the imperial domes of Mount Washington and the adjacent mountains, complete the framing of this valley, the Mecca of thousands of pilgrims every year. Here is the beautiful station of the Conway Division of the Eastern Railroad, where we may take an express train direct for Boston, via Great Falls and Portsmouth, with Pullman cars. The Portland and Ogdensburg Road also has a station near by, whence those who desire may proceed to Portland and the East.

THE EPILOGUE.

NOTHING now remains to be said but the word "Farewell," — a word sad at all times, but how inexpressibly mournful when the occasion is the separation after a vacation journey, with all its joys and vexations, sublime scenery, and dusty railroad rides, romantic climbs and torn clothes, inspiration of Nature and botheration of baggage, sweets of flirtation and bitters of bill-paying — yet there is this consolation in our separation, that next season our experience can be renewed, if we choose, for the attractions described are perennial, our little book is always ready and always reliable, and only our own choice is needed to revisit some of the bright scenes that we have enjoyed together by the way.

Having sipped the Saratoga waters together, sailed up and down the lovely lakes together, climbed in and out of the chasm and over the Adirondack Mountains together, plunged over the Falls of Niagara together (figuratively, as it were), made the tour of the St. Lawrence, the Saguenay and the White Hills together, it will be strange indeed, if we do not find ourselves with each recurring season, taking as a part of our vacation enjoyment and recuperation, a trip to some of the resorts which we have just visited.

A Pleasant Excursion.

A very pleasant trip may be made from Boston by taking either of the line steamers of

SANFORD'S INDEPENDENT LINE TO BANGOR.

THE CAMBRIDGE, CAPT. J. P. JOHNSON, OR THE KATAHDIN, CAPT. W. R. ROIX.

The boats of this line leave Boston alternately,

Every Monday, Tuesday, Thursday and Friday, at 5 P. M.

The tourist desiring to visit that famous resort, Mount Desert, finds the easiest and speediest connection, via Steamer Lewiston, at Rockland.
"Mount Desert" is the name of an island of about a hundred square miles in extent, which enjoys the highest reputation among the many places of resort upon that coast. It has been described as the most extraordinary island on the coast of Maine, and perhaps on the whole coast of America. It is remarkable for its size, its bold and wild scenery, its pure air, and is not without a curious history. The name formerly belonged to a township which comprised the whole island; it is now retained by the central portion; the northerly portion is a township incorporated under the name of "Eden;" and the southerly portion under the name of "Tremont." The appearance of the island when approached from the sea is grand and romantic in the extreme. In the central and south-eastern parts, the island maintains its native wildness, and game abounds in great variety. The great attraction of the place is found in the combination of the charms of ocean and mountain scenery, both of which may be enjoyed here in the highest degree. Whether for trout fishing, or for deep-sea fishing, the facilities offered at Mount Desert are unsurpassed elsewhere. The principal mountain upon the island has been accurately ascertained by the United States Coast Survey to be 1480 feet above the level of the sea. This is an extraordinary altitude for an eminence within a few miles of the deepest waters of the ocean. It is a favorite place for the sojourn of artists; and it is said that whoever visits it once, is sure to return again and again. The air is especially pure and bracing; it is strengthening to invalids, and is breathed with exquisite pleasure by those in full health. Many persons from the cities, after becoming acquainted with the beauties of the island in repeated visits, have built houses there in order to make it a place of permanent residence during the summer months. There are several centres of attraction upon the island. The principal Post-office address is "Bar Harbor." The island is practically almost inaccessible to travellers by any land route. The best method of reaching it is by the Steamers of Sanford's Independent Line from Boston to Bangor. One of these Steamers leaves Boston at five o'clock in the afternoon, every Tuesday and Friday; passengers for Mount Desert are transferred at Rockland to the Steamer Lewiston, by which they reach their destination without delay.

The Wassaumkeag House, at Fort Point, on the Penobscot River, is another favorite resort reached by this line, and visitors to Moosehead Lake, Sebec, St. John, and other points "down East," will get rapidly and comfortably to their destinations. The sea voyage gives just enough variety to spice the otherwise rather monotonous existence of travel, and the comforts to be found on board make the tourist feel at home all the way.

FIRST-CLASS HOTELS.

AMERICAN HOUSE..........................St. Albans, Vt.
    H. PIERCE & SON, Proprietors.
AMERICAN HOUSE..........................Montpelier, Vt.
    CHESTER CLARK, Proprietor.
AMERICAN HOUSE..........................Burlington, Vt.
    H. H. HOWE, Proprietor.
AMERICAN HOUSE..........................Fitchburg, Mass.
    E. DEWOLF & CO., Proprietors.
AMERICAN HOTEL..................Saratoga Springs, N. Y.
    WM. BENNETT, Proprietor.
BISHOP'S HOTEL..........................Montpelier, Vt.
    H. H. BISHOP, Proprietor.
BERWICK HOTEL............................Rutland, Vt.
    C. F. RICHARDSON, Proprietor.
BRATTLEBORO' HOUSE.....................Brattleboro', Vt.
    H. A. KILBURN, Manager.
CONGRESS HALL............................Albany, N. Y.
    ADAM BLAKE, Proprietor.
CONGRESS HALL...................Saratoga Springs, N. Y.
    HATHORN & SOUTHGATE, Proprietors.
CLARENDON HOTEL.................Saratoga Springs, N. Y.
    C. E. LELAND, Proprietor.
COLUMBIAN HOTEL.................Saratoga Springs, N. Y.
    D. A. DODGE, Proprietor.
CLIFTON HOUSE.....................Niagara Falls, N. Y.
    COLBURN & MCOMBER, Proprietors.
CRAWFORD HOUSE...........................Carroll, N. H.
    A. T. & O. F. BARRON, Proprietors.
FT. WILLIAM HENRY HOTEL..Lake George (Caldwell), N.Y.
    T. ROESSLE & SON, Proprietors.

HOTELS—Continued.

FOUQUET'S HOTEL .... Plattsburg, N. Y.
    SMITH & MARTIN, Proprietors.
GRAND UNION HOTEL .... Saratoga Springs, N. Y.
    J. H. BRISLIN & CO., Managers.
HOLDEN HOUSE .... Saratoga Springs, N. Y
    C. H. HOLDEN, Proprietor.
INTERNATIONAL HOTEL .... Niagara Falls, N. Y.
MASON'S HOTEL .... North Conway, N. H.
    F. H. MASON, Proprietor.
MARTIN'S HOTEL .... Saranac Lake, Adirondacks, N. Y.
    W. F. MARTIN, Proprietor.
MONTEAGLE HOTEL .... Niagara Falls, N. Y.
    ALEXANDER & TERRILL, Proprietors.
MAGOG HOUSE .... Sherbrooke, P. Q.
    F. P. BUCK, Proprietor.
ALARRIN HOUSE .... Saratoga Springs, N. Y.
    H. A. QUACKENBUSH & CO., Proprietors.
OTTAWA HOTEL .... Montreal, P. Q.
    BROWNE & PERLEY, Proprietors.
PAVILION HOTEL .... Wolfeborough, N. H.
    A. L. HOWE, Proprietor.
PEMIGEWASSET HOUSE .... Plymouth, N. H.
    C. M. MORSE, Proprietor.
RUSSELL HOUSE .... East Milton, Mass.
    JAMES M. RUSSELL, Proprietor.
ROCKWELL HOUSE .... Glens Falls, N. Y.
    ROCKWELL BROTHERS, Proprietors.
STEVENS HOUSE .... Vergennes, Vt.
    S. S. GAINES, Proprietor.
ST. REGIS LAKE HOUSE .. St. Regis Lake, Adirondacks, N. Y.
    PAUL SMITH, Proprietor.
ST. LOUIS HOTEL .... Quebec, P. Q.
    W. RUSSELL & SON, Proprietors.
TWIN MOUNTAIN HOUSE .... Carroll, N H.
    A. T. & O. F. BARRON, Proprietors.
UNITED STATES HOTEL .... Saratoga Springs, N. Y.
    Hon. JAMES M. MARVIN, Proprietor.
UNION HOTEL .... Cuttingsville, Vt.
    H. TODD, Proprietor.
VAN NESS HOUSE .... Burlington, Vt.
    BARBER & FERGUSON, Proprietors.
WELDON HOUSE .... St. Albans, Vt.
    THOMAS LAVENDER, Proprietor.
WAVERLY HOUSE .... Saratoga Springs, N. Y.
    Major W. J. RIGGS, Proprietor.

INDEX TO ADVERTISEMENTS.

Lake Shore and Michigan Southern Railway, . . 4
Adriondack Company's Railroad, ... 5
Chicago, Milwaukee & St. Paul Railway, . . . 6
Michigan Central & Great Western Railway, . . 7
Montreal & Boston Air Line, ... 8
St. Lawrence & Saguenay Line, ... 9
Quebec & Gulf Ports Steamship Co., ... 10
Eastern & Maine Central Line, ... 11
Chicago, Burlington & Quincy Railroad, . . . 12
Congress and Empire Springs, ... 13
Ottawa Hotel, ... 14
Monteagle Hotel, ... 15
St. Louis Hotel, ... 16
Eagle Hotel, ... 17
Brattleboro House, ... 18
Bellevue House, ... 19
Magog House, ... 20
Congress Hall, }
Rollstone House, } 21
Boston Daily Globe, ... 22
Boston, Concord, Montreal & White Mountains R R., 23
Bray & Hayes, }
Boston Courier, } 24

Lake Shore and Michigan Southern Railway.

The Great Double Track Route
From BOSTON and NEW YORK
To CHICAGO, via BUFFALO.

The only Line connecting with the
NEW YORK CENTRAL AND ERIE RAILWAYS
Running Through Cars without Transfer of Baggage.

Six Express Trains leave Buffalo
—DAILY FOR—
CLEVELAND, TOLEDO, CHICAGO AND ST. LOUIS,
WITHOUT CHANGE.

A *Lake Shore Palace Sleeping Car runs* between *Niagara Falls and Chicago,* via *East Buffalo* (daily, Sundays excepted), leaving Niagara Falls 11.33 A. M., and accompanies the Chicago train leaving East Buffalo, 12.15 P. M., arriving Chicago, 8.20 A. M.

Only one Change of Cars between BUFFALO and ST. JOSEPH, KANSAS CITY, LEAVENWORTH, OMAHA and ALL POINTS WEST and SOUTHWEST.

SECURE TICKETS BY THIS FAVORITE ROUTE.
For Sale at all Principal Offices in the East.

Sections secured in Wagner Cars, Boston to Chicago, at the Company's Office, No. 210 Washington St., (Old State House,) and on application to Wagner's Agents.

JAS. S. SMITH, Agent,   J. A. BURCH, Gen'l Eastern Agent,
   210 Washington St., Boston.   *Buffalo, N. Y.*
         CHAS. PAINE, General Sup't, CLEVELAND, OHIO.

W. W. RUGGLES, Gen'l Trav. Agent for Mass. and Northern New England, Boston, Mass.

W. A. CROMWELL, Gen'l Trav. Ag't for East'n New England and Provinces, Boston, Mass.

Adirondack Company's Railroad.

FROM SARATOGA SPRINGS,
—TO—
HADLEY, (LUZERNE,) THURMAN, (Station for WARRENSBURG and LAKE GEORGE,) RIVERSIDE, AND NORTH CREEK,

FORMING THE

MOST DIRECT RAILROAD ROUTE

—TO THE—

VALLEY OF THE UPPER HUDSON,
And the Wilderness.

Express Trains leave Saratoga Springs on arrival of morning and mid-day (N. Y. Special) trains from the south.

Connections are made at Thurman with a First-Class Stage Line to Lake George.

The distance by Stage, (9 miles), Through Fare and Time being the same as by the old route via Glen's Falls. This route affords

New and Far More Picturesque and Delightful Scenery

Than any other route from Saratoga.

Also, at RIVERSIDE with Stages for CHESTER, POTTERSVILLE, and the STEAMBOAT on SCHROON LAKE.

Also, at NORTH CREEK (from morning Train) with Stages for WARBURN'S, (*Indian Lake,*) JACKSON'S, (*Cedar River,*) WAKELEY'S (*Cedar Falls,*) and the new Hotel at BLUE MOUNTAIN LAKE, 29 miles distant, the most desirable rendezvous and starting point from which to reach RAQUETTE LAKE, and the HEART OF THE GREAT FOREST.

C. E. DURKEE,
General Ticket Agent.

C. H. BALLARD,
Superintendent.

Chicago, Milwaukee & St. Paul
RAILWAY.

THE GREAT THROUGH LINE BETWEEN

Chicago, New York, New England, the Canadas,

———AND———

All Eastern and Southern Points,

AND THE GREAT NORTHWEST.

Connecting in Chicago with all Eastern and Southern Lines.

CHICAGO DEPOT:—Corner Canal and West Madison Sts. Horse Cars and Stage Lines for all parts of the City constantly passing.

CHICAGO CITY OFFICES:—61 and 63 Clark Street.

MILWAUKEE DEPOT:—Corner Reed and South Water Sts. Horse Cars and Omnibus Lines running regularly therefrom to the principal parts of the City.

CITY TICKET OFFICE:—400 East Water St. cor. Wisconsin Street.

THE ONLY THROUGH LINE BETWEEN
Chicago, Milwaukee, St. Paul and Minneapolis.

It traverses a finer country, with grander scenery, and passes through more business centres and pleasure resorts, than any other Northwestern Line.

The Only Railway Line along the Valley of
THE UPPER MISSISSIPPI RIVER,
AND THE SHORE OF LAKE PEPIN,

Also via Madison, Prairie du Chien, McGregor, Austin and Owatonna.

THROUGH PALACE COACHES AND SLEEPING CARS OF THE BEST,
AND TRACK PERFECT.

☞ Connecting at St. Paul and Minneapolis, with the several lines centering at those points.

ST. PAUL DEPOT:—Corner of Jackson and Levee Streets.

CITY OFFICE:—118 East Jackson Street, corner of Third Street.

A. V. H. CARPENTER, Gen. Pass. and Ticket Agent, Milwaukee.

BOSTON OFFICE: 1 Court St., E. L. HILL, Agent.

Michigan Central and Great Western Railways,

VIA

BUFFALO, SUSPENSION BRIDGE,

AND

Niagara Falls.

4 Through Express Trains Daily to Chicago.

Pullman and Wagner's luxurious Drawing-Room, Hotel and Palace Sleeping Cars run on all Through Express Trains over this Line, with

ONLY ONE CHANGE OF CARS FROM BOSTON TO CHICAGO.

THIS IS THE

SHORTEST, QUICKEST AND MOST DESIRABLE LINE

BETWEEN THE

NEW ENGLAND STATES

AND

CHICAGO, MILWAUKEE, ST. PAUL,

AND THE PACIFIC COAST.

Passengers purchasing their tickets by this route are allowed to stop off and resume their journey at pleasure, thus affording them an opportunity of witnessing the Greatest Natural wonder in America, the

FALLS AND SCENERY OF NIAGARA.

Baggage Checked Through to all Points West.

Be sure and ask for Tickets via

The Great Western and Michigan Central Railways,

Which are sold at all principal offices east of Suspension Bridge.

A. J. HARLOW, **J. Q. A. BEAN,**
Eastern Passenger Agent,   Gen'l Eastern Agent,
201 Washington St., Boston.   349 Broadway, New York.

*THE GREAT NORTHERN ROUTE FOR TOURISTS
AND PLEASURE SEEKERS.*

THE NEW

Montreal and Boston Air Line,

—COMPOSED OF THE—

Boston, Concord & Montreal R. R., Concord to Wells River,
Passumpsic R. R., Wells River to Newport, Vt.,
South-Eastern Railway, Newport to St. Johns, P. Q.

Will run 2 FAST EXPRESS TRAINS,

Composed of NEW AND ELEGANT CARS provided with
all modern improvements,

From BOSTON (LOWELL DEPOT,) to MONTREAL,

Where connection is made with Grand Trunk Railway for the West.

———o———

Entire Trains, with Pullman Cars attached, run from Boston to Montreal without change, and only one change to Chicago. No route from Boston presents such magnificent scenery, and Passengers by this Line travel through the *Paradise of this Continent.*

A continuous and most charming Panorama of River, Mountain, Valley and Lake Scenery will entertain the traveler for a distance of 250 miles, including the grand views of

LAKE WINNIPESAUKEE,
THE WHITE MOUNTAIN RANGE,
PASSUMPSIC RIVER VALLEY,
CRYSTAL LAKE,

AND THE

ROMANTIC LAKE MEMPHREMAGOG.

———o———

Trains stop 30 minutes for meals at the Pemigewasset House, Plymouth, N. H., and the Memphremagog House, Newport, Vt.

GENERAL OFFICE,

240 WASHINGTON STREET,

(Old Number 94.]

GUST. LEVE, Gen'l Agent.

THE St. Lawrence & Saguenay
LINE OF STEAMERS,
PLYING BETWEEN

QUEBEC,

THE

RIVER SAGUENAY,

AND THE

Watering Places of the lower St. Lawrence,

IS COMPOSED OF THE

First-Class Sea-Going Passenger Steamers,

"SAGUENAY,"

"ST. LAWRENCE,"

and "UNION."

ACCOMMODATIONS FIRST-CLASS.
CHARGES MODERATE.

Tickets for sale at all Principal Ticket Offices in the States and Canada; and at the Office of the Company, opposite ST. LOUIS HOTEL, QUEBEC, or ST. ANDREWS WHARF.

For further information apply to

STEVENSON & LEVE, General Agents,　　**A. CABOURY,**
　　BOSTON: 240 Washington Street.　　　　SECRETARY.
　　MONTREAL: 202 St. James Street.

A GREAT ATTRACTION

To Tourists and Pleasure Travelers,

IS THE ROUTE OF THE

Quebec and Gulf Ports

STEAMSHIP CO.

Whose commodious Steamers proceed from Quebec down the majestic River and Gulf of St. Lawrence, in sight of the grandest scenery and many historical points, calling at numerous noted sea-bathing resorts on the south shore of the Gulf, giving the sportsman and angler a chance to visit the most far-famed rivers, bays and inlets, which swarm with trout and salmon.

The Steamers connect at Point du Chene (Shediac) with the Intercolonial Railroad for St. John, N. B., thence by cars and steamers to

PORTLAND AND BOSTON,

and at Pictou with Intercolonial Railroad for Halifax, N. S., connecting there with Railroad or Steamer lines for St. John, Portland and Boston. This is the route to CHARLOTTETOWN and PRINCE EDWARD ISLAND.

Excursion Tickets,

From New York, Boston, or other points in New England, to Montreal, Quebec, thence via Gulf Port Steamers to Shediac, N. B., Charlottetown, P. E. I., Pictou, N. S., St. John, N. B., Halifax, N. S., passing through all points of interest in the maritime provinces, and returning by either rail or steamer to Portland, Boston and New York; or vice versa.

For sale in New York, Boston, and principal points in New England, AT ALL OFFICES SELLING EXCURSION TICKETS.

Ask Ticket Agent for Gulf Ports Steamer circular, which will give you all the particular information, and map of route.

STEVENSON & LEVE, Passenger Ag'ts.   W. MOORE, Manager.

GENERAL OFFICE, Opposite St. Louis Hotel, QUEBEC.

Montreal, -   - 202 St. James St.

BOSTON, - - - 240 (old number, 94) Washington Street.

Fo
Plac

This fold-out is being

Eastern & Maine Central
RAILROAD LINE.
THE GREAT THROUGH ROUTE
TO THE
State of Maine & Maritime Provinces,
ALSO THE

SHORTEST and popular route via the Sea Shore, Hampton and Rye Beaches, and Isles of Shoals, to Wolfboro, North Conway, and White Mountains.

The only direct route to the **RANGELEYS** and **MOOSEHEAD LAKE**, the **GREAT FISHING RESORTS**.

CONNECTIONS are also made at Portland with the Grand Trunk Railway, for Gorham, the Canadas, and the West, also with all the Steamboat Lines to Mount Desert, the coast of Maine and the Maritime Provinces.

NO CHANGE OF CARS
Between Boston and North Conway, or Boston and Bangor, and but one to St. John, N. B.

PULLMAN PALACE CARS are in use on this Line.

No other line offers such facilities or possesses such advantages, to the great pleasure resorts of New England. All the modern improvements are in use on this road.

SEATS or BERTHS IN PULLMAN CARS
Can be secured by letter or telegraph, at the

Boston Office, 280 Washington Street.

Before purchasing Tickets refer to Maps, Advertisements, etc., of this Company, to be obtained at the Ticket Offices in New York, Philadelphia, Baltimore, Washington, Montreal, Quebec, and White Mountains, also of the principal Ticket Agents in the United States and Maritime Provinces.

CHAS. F. HATCH, Gen'l Manager.
GEO. BACHELDER, Sup't E. R. R.   PAYSON TUCKER, Sup't M. C. R. R.
GEO. F. FIELD, Gen'l Pass. Agent.

THIS IS AN ADVERTISEMENT

AND IS PAID FOR AS SUCH,

| BUT NOTWITHSTANDING |

It is literally true, as thousands can and will testify, that the

Chicago, Burlington and Quincy
RAILROAD

Has the SMOOTHEST AND BEST TRACK, and the BEST AND MOST COMPLETE EQUIPMENT of all roads in the West, and has NO SUPERIOR in any part of the country.

It is the Favorite with the Traveling Public.

It is the only line to

CALIFORNIA

Running the justly Celebrated and most Comfortable

DINING CARS

And offers the very best route to all points in

**KANSAS, COLORADO,
NEW MEXICO, IOWA,
NEBRASKA, MISSOURI AND TEXAS.**

No Passenger will ever regret having chosen this Route.

TICKETS via this Line FOR SALE AT ALL THE TICKET OFFICES IN THE EAST.

COMPANY'S OFFICES:
317 Broadway, New York. 222 Washington St., Boston.

D. W. HITCHCOCK,   E. P. RIPLEY,
Gen'l Pass. Agent, CHICAGO.   Gen'l Eastern Pass. Agent, BOSTON.

1790. | **SARATOGA SPRINGS, NEW YORK.** | **1875.**

THE

CONGRESS and EMPIRE

SPRING WATERS,

Are the Best Anti-Bilious Remedies known.

---o---

They are purely natural mineral waters, cathartic, alterative and slightly stimulating and tonic in their effects, without producing the debility that usually attends a course of medicine.

They are used with marked success in Affections of the Liver and Kidneys; and for Dyspepsia, Gout, Chronic Constipation, and Cutaneous Diseases they are unrivalled.

They are especially beneficial as general preservatives of the tone of the stomach and purity of the blood, and are powerful preventatives of Fevers and Bilious Complaints.

These waters should be taken in the morning before breakfast—one pint being the usual draught—and their use may be continued daily for months, with the most agreeable results, and without reaction, or any necessity of increasing the quantity taken.

CONGRESS WATER being the most popular of the Saratoga waters, is largely counterfeited. Every genuine bottle has the letter *"C,"* with the name of the undersigned Proprietors, and the words CONGRESS WATER raised upon the glass.

The genuine EMPIRE WATER may be similarly distinguished, the letter *"E,"* and words EMPIRE WATER, being substituted for those above named.

Purchasers should require these brands and marks.

☞ *None Genuine sold on draught.*

Orders by mail receive prompt attention. Address

CONGRESS & EMPIRE SPRING CO.,

94 CHAMBERS STREET,

Or, SARATOGA SPRINGS, N. Y. *NEW YORK.*

MONTREAL.

C. S. BROWNE, J. Q. PERLEY, Proprietors.

THIS well known and popular Hotel is situated on St. James Street, the principal business street of the city, and is near the Post Office, Banks, Theatre, and all the Public Buildings, and has ample accommodation for 400 guests.

The Ottawa Hotel covers the entire space of ground running between St. James and Notre Dame Streets, and has two beautiful fronts. The House has been thoroughly Refitted and Furnished with every regard to comfort and Luxury—has hot and cold water, with baths and closets on each floor. The aim has been to make this the

Most Unexceptional First-Class Hotel in Montreal.

The proprietors respectfully assure their patrons that no exertions will be spared to make this Hotel

A Comfortable Home for the Traveling Public.

Carriages, with attentive drivers, may be had at all times by application at the office.

Coaches will be found at the Railway Depot and Steamboat Landings, on the arrival of the several Trains and Steamers.

Montreal Telegraph Office in the House.

Monteagle House,

SUSPENSION BRIDGE, NIAGARA, N. Y.

This House is now open for the reception of its patrons and the traveling public, under an entire new management. Having been refitted and entirely renovated it commands the attention of parties visiting Niagara. The rooms command a fine, uninterrupted view of Niagara Falls, the two Suspension Bridges, Whirlpool, and Whirlpool Rapids. In connection with the house are Mineral Sulpher Springs and Baths, making it desirable for those requiring tonic and cutaneous treatment.

Free Omnibus to and from all trains.

Terms, $3.00 per day.

Special inducements to parties remaining any length of time. Parties intending to stop at the Monteagle, should have their baggage checked to Suspension Bridge.

ALEXANDER & TERRILL, Proprietors.

ST. LOUIS STREET,

QUEBEC.

WILLIS RUSSELL, - - - - Proprietor.

This Hotel, which is unrivalled for size, style and locality, in Quebec, is open through the year for pleasure and business travel.

It is eligibly situated near to, and surrounded by the most delightful and fashionable promenades,—the Governor's Garden, the Citadel, the Esplanade, the Place d'Armes, and Durham Terrace—which furnish the splendid views and magnificent scenery for which Quebec is so justly celebrated, and which is unsurpassed in any part of the world.

The proprietor, in returning thanks for the very liberal patronage hitherto enjoyed, informs the public that this hotel has been enlarged and refitted, and can now accommodate five hundred visitors; and assures them that nothing will be wanting on his part that will conduce to the comfort and enjoyment of his guests.

EAGLE HOTEL,

OPPOSITE THE CAPITOL GROUNDS,

CONCORD, NEW HAMPSHIRE.

Enlarged, Refitted, and Refurnished.

TOURISTS

Wishing to make a stop on the way to or from the Mountains will find accommodations equal to any in New England.

FREE CARRIAGES TO AND FROM THE DEPOT,

JOHN. A. WHITE, Proprietor

BRATTLEBORO HOUSE,
BRATTLEBORO, VT.

H. A. KILBURN, Manager.

BELLEVUE HOUSE

H. BEAN, - - - - PROPRIETOR.
NEWPORT, VT.

The Bellevue was constructed last season, and opened for the accommodation of the public Dec. 1st, 1873. It has all of the modern improvements, lighted by gas, large pleasant rooms, and for neatness and conveniences is unsurpassed by any house in the State. Being situated at the head of Lake Memphremagog, and at the junction of the S. E. Railway with the C. & P. R. & M. V. R. R., makes it a very desirable and accessable point both for commercial and pleasure travellers.

The management, grateful to the public for the favors extended to him for the past ten years, solicits a continuance of their patronage, with the assurance that the Bellevue shall be second to none in its comforts and accommodations.

RATES PER DAY:—For Commercial Travellers, $2.00; for Pleasure Visitors, $2.50,

☞ Livery furnished for guests at reasonable rates.

NEWPORT, June 1st, 1875.

MAGOG HOUSE
SHERBROOKE, P. Q.

At Newport, Vt., the traveller can take the elegant Steamer

"LADY OF THE LAKE,"

CAPT. FOGG,

To Magog, and from that point, John Norton's Stage Line to Sherbrooke, where they will find the

| MAGOG HOUSE, |

Under the management of Mr. H. S. HELPBURN, one of the best kept Houses in Canada.

Excellent Fishing in the Vicinity.

The far famed Lakes

MASSAWIPPI AND MEGANTIC

Are but a short distance from the House.

The Steamer, the Stage Line, and the House are under the control of men who from long experience are well qualified to anticipate and satisfy the wishes of travellers.

CONGRESS HALL
ALBANY, N. Y.

Situated on Capitol Hill, adjoining the New Capitol, and State Library; fronting on the City Parks, and in the immediate vicinity of all the State Buildings.

Fine Cuisine, Large and Airy Rooms,
WITH ALL THE MODERN IMPROVEMENTS.

Pleasure seekers will find this the most pleasant Hotel to stop at in the City.

ADAM BLAKE, Proprietor.

| FIRST-CLASS HOTEL. |

Rollstone House,
FITCHBURG, MASS.

Wm. F. Day, - - Proprietor.

Formerly of FITCHBURG HOTEL and AMERICAN HOUSE.

Guests conveyed to and from the Depots free of charge.

BOSTON

DAILY GLOBE,

FOLDED, PASTED AND CUT

LIKE A BOOK.

PRICE THREE CENTS.

Boston, Concord, Montreal
AND
WHITE MOUNTAINS R. R.

Now open to the Fabyan House, White Mountains, and to Northumberland on the Grand Trunk Railway.

The Shortest, Quickest and Best Route
TO THE
White and Franconia Mountains, Montreal and Quebec.

This is the only line running Day Palace Cars and Express Trains
between Boston, Providence, Worcester, New London, Stonington, and the White and Franconia Mountains. This
Line passing as it does up the valley of the Merrimac
and Connecticut Rivers, through the Cities of
Lowell, Lawrence, Nashua, Manchester, and
Concord, and along the borders of Lake
Winnepesaukee for 30 miles, terminating at the foot of Mt. Washington, passing River, Lake, and
Mountain scenery unequalled in New
England, and in the immediate vicinity of the
principal Hotels and summer Boarding Houses
in the Northern part of the State, makes it
the popular travellers route for
Tourists visiting the

Lake and Mountain Scenery of New Hampshire.

For further information regarding time, connections, tickets, &c., see the principal R. R. Guides, or apply to

No. 5 State Street, or 240 Washington Street, Boston,
And the principal Offices on the Line.

J. A. DODGE, Sup't, Plymouth, N. H.

BRAY & HAYES,

99 and 101 Broad Street, Boston,

Importers and Commission Agents,

S. Rae & Co.'s Italian Oils. Keen's English Mustard.

French Prunes, Sardines, Salad Oil, Olives, Mushrooms, Truffles, French Peas, Capers, China Preserved Ginger, &c., True Castile Soap, White and Mottled, Fine Toilet Soaps, English Pickles, Sauces and Catsups, Worcestershire Sauce, Mustard (tins and bottles), India Soy, and Currie Powder, Potted Ham, Anchovies, &c. Liebie's Food for Infants, Groats and Barley for Infants, London and Dublin Porter, and Scotch Ale, Scotch Oatmeal (lbs. and bbls), Cox's Gelatine ; French Gelatine, Marmalades, Jellies, &c., Edinburg Albert Biscuit, London Biscuits (tins and bbls.), Liebig's Extract of Beef, Italian Macaroni, &c., Chamois Skins, &c.

DAY & MARTIN'S LONDON BLACKING. MAY'S ENGLISH WASHING POWDER.
CROSSE & BLACKWELL'S ENGLISH PICKLES AND CONDIMENTS.

1824. | THE BOSTON COURIER FOR 1875. | 1875.

This widely known paper has been established for half a century, and it is believed it has maintained, during this long period, a reputation unsurpassed by any other in the United States. Within the past year, and since its enlargement and the introduction of several new and special features, THE COURIER has made extraordinary advances in public favor, and its subscribers and readers have increased to an extent which its proprietors believe altogether unparalleled, for so short a time, in the records of a newspaper enterprise in New England. THE COURIER is a paper to be read, and its proprietors have ample means of knowing that it is carefully perused by great numbers of all classes of persons throughout the range of its circulation, and is much sought for by many leading minds in all departments of life.

It is obvious how important a medium of advertising such a paper must be, a fact of which we exhibit plenty of evidence that the intelligent public is conscious; and taking advantage of this opportunity to thank their numerous advertisers for past favors, the proprietors of THE COURIER would respectfully solicit from their friends and the public the patronage which they may think justly due to this long-established and well-known paper. The favor with which it is received warrants us in every effort to promote its circulation, and full reliance may be felt, that it will maintain its character, and all pains will constantly be taken to make the paper in every respect useful and entertaining.

TERMS—One copy by mail, one year, $2.75 ; by carrier, one year, $3.00.
Payment to be made invariably in advance.

JOSEPH F. TRAVERS, Publisher,
299 Washington Street, - - - - - BOSTON.

LAKE VIEW HOTEL,
WATKINS, N. Y.

OPEN FOR GUESTS FROM JUNE 1st UNTIL NOVEMBER 1st.

Beautifully situated on an eminence overlooking the village and SENECA LAKE, and but five minutes' walk from the wonderful world-renowned WATKINS GLEN, one of the most romantic spots on the American Continent.

This House is replete with every comfort and luxury demanded by the modern tourist, and will accommodate 150 guests. It challenges comparison with any summer resort hotel in the country—large rooms, high ceilings and thorough ventilation—pure spring water and gas throughout—hot and cold baths—telegraph office in the house—extensive grounds with groves and evergreen forest trees, and scenic attractions unsurpassed in the country.

W. T. PURVIANCE.

J. M. BETTMAN, Manager.